AF386605

Justified Killing

Stephen Napier

Justified Killing

An Essay on Interests, Persons, and Epistemic Justification

 Springer

Stephen Napier
Department of Philosophy
Villanova University
Villanova, PA, USA

ISBN 978-3-032-14945-9   ISBN 978-3-032-14946-6   (eBook)
https://doi.org/10.1007/978-3-032-14946-6

Philosophy and Medicine Series

For Harvey, Carol, Katherine and VFL
"Si quis sitit, veniat ad me et bibat.
Qui credit in me,…flumina de ventre ejus
fluent aquae vivae."

Competing Interests The author has no competing interests to declare that are relevant to the content of this manuscript.

Contents

Chapter 1
Introduction: *De Jure* Objections and the Justification for Killing

1.1 Introduction

Betty was a 19-year-old woman who'd been dating a man for quite some time. They talked about marriage, about having children. Eventually she became pregnant and shared this news with her boyfriend expecting some level of acceptance and joy. To her surprise and shock, her boyfriend expressed frustration, told her that it's her problem, and that she needs to get an abortion. Feeling trapped, abandoned, and scared. Her boyfriend threatened to leave her if she did not get an abortion. Lacking a job and any other resources she knew she had to get an abortion.[1] Betty was not thinking about morals, personhood, or other philosophical technicalities. She could only think that she could not have a child in such circumstances. She was thinking in terms of what could she do? What would be best for her right now?

Jerry is a 46-year-old male who had been diagnosed with Huntington's disease 6 years prior. He is now experiencing severe chorea, myoclonus, dysphonia (though he is partially able to communicate via writing). He requires PEG feeding due to his inability to swallow. His psychiatric status is growing more dismal, including irritability, depression, hopelessness, and resignation. He retains executive functioning and his "insight" on his disease state is preserved—making his daily experience even more existentially painful. Treatment for the existential issues is diminishing in efficacy while the chorea and myoclonus require heavy sedatives. In his wakeful moments, he requests to be euthanized describing himself as trapped in a body and mind that never lets him rest.

Betty faces a "crisis" pregnancy, and Jerry faces real anxieties about dying. Their fears, suffering, and desire to do what is best are real, and their circumstances are unquestionably difficult. Nothing that is said in this book aims to diminish or ignore

---

[1] This story is based loosely on the story of Betty in the documentary *When Abortion Was Illegal.* Bullfrog Films. Retrieved from https://video.alexanderstreet.com/watch/when-abortion-was-vvillegal

S. Napier, *Justified Killing*, https://doi.org/10.1007/978-3-032-14946-6_1

the reality of what those like Betty or Jerry might feel, and the temptations they might face. What is argued here, however, is that there is a boundary for acceptable solutions to such problems. That line is drawn at intentional killing. Stating the idea this way, though succinct, ignores considerable nuance. The following reflections fill in some details.

1.2 Justification and Risk

Suppose you are out for a hike with your 10-year-old daughter and happen upon a small body of water. You are both aquaphiles and want to go for a swim. You hesitate, though, not sure of how deep the water hole is. You also don't know what is in it—suppose it's a watering hole along the Amazon river. You don't know it is safe, and the risk of harm does not outweigh the joy of swimming. The belief that it is safe inherits a risk in being wrong. Now suppose that to be sure that it is safe, you propose that your daughter jumps in first? What moral adjectives would you use to describe a father who did such a thing?

The case illustrates two ideas. First, some of our decisions have a cost in being wrong, specifically a moral cost in that if one is wrong that the action is permissible, one performs an immoral action— jumping in hastily and harming oneself. The example also illustrates that for some of our decisions, the consequences of being wrong are born by others. Being wrong that the water hole is safe to swim in has consequences for my daughter, not me. Likewise, if a clinician is wrong that abortion or euthanasia are permissible, somebody else is killed.

The topics of this work are the arguments for permissible killing in bioethics, i.e., abortion and euthanasia. Specifically, the task is to navigate how the justifications for and against these actions relate to the notion of moral risk. In light of this task, the project focuses on two concepts, justification and moral risk. What are these two notions?

It is common in epistemology to distinguish between the epistemic concepts of truth and justification. It is also common to suppose that one's beliefs can be justified and yet some or all of her beliefs are false. Conversely, we can imagine an epistemic agent who believes many truths but does not have any or very poor justification for them. For example, suppose an agent uncritically accepts the testimony of others and it just so happens that every testimonial report is true. So, analyses of whether our beliefs are true can be considered somewhat separate from whether they are justified. Alvin Plantinga (2000) distinguishes between these projects as making de facto (i.e. about the facts) claims versus de jure (i.e. about the justification) claims. Likewise, one can make de facto objections to an argument according to which one challenges the truth of the claims being made. And one can make de jure objections to an argument according to which one challenges the justification for those claims. So, there are two ways to object to a belief: object to its truth, and object to its justification or rationality.

This project focuses exclusively on de jure challenges to arguments for the permissibility of killing (hereafter (pk))— specifically arguments on abortion and euthanasia. So, the first nuance is that even if such judgments are true, I argue that the typical arguments made in support of permissible killing suffer from serious structural flaws; i.e., they fall prey to de jure objections. Which flaw the argument suffers from depends upon the specific argument in question.

But if there are de jure flaws, why introduce the notion of moral risk? What does that concept add to the story told here? A full answer to this question awaits Chap. 8, but the basic idea is this. Suppose Jones thinks that abortion is permissible and that none of the pro-life arguments have convinced him otherwise. Jones enjoys a coherent network of mutually supporting beliefs about abortion and related topics. But now suppose that someone articulates that the support relations Jones took for granted do *not* entail that abortion is permissible. That is, suppose there is a good argument to the effect that Jones' beliefs on these issues look less like a net of interconnected and woven nodules, and more like pieces of rope in the same vicinity of each other like a dirempted cargo net. Exploiting a different metaphor, Jones' beliefs are collocated but not connected; they are vulnerable to de jure objections.

How should Jones respond? One response is to say that the pro-life arguments are just as bad, and that though his beliefs lack some threshold of interconnectivity or justification, they are still true. The epistemic relevance of moral risk allows for this response but argues that Jones would not be justified *in acting* on his belief that abortion or euthanasia are permissible. This is the second notable nuance. If his beliefs look more like dirempted pieces of rope than a net and given the moral cost of being wrong, Jones should not act. That is the conclusion of this work, but it requires arguing both that the justifications for (pk) suffer de jure defects *and* that (pk) bears a heavier or more serious moral risk in being wrong than ~(pk). The claim (pk) may be true, but one would not be justified in acting on it.

There is no doubt that this conclusion is heterodox in relation to the broader academy. There are received opinions in contemporary bioethics on issues such as abortion and euthanasia. A recent survey of bioethicists indicates a rather homogenous view of the profession according to which an overwhelming majority think that abortion (87%) and, for a slightly smaller majority, euthanasia are permissible. In fact, the only other issue that garnered narrowly more agreement on "abortion is permissible" was in response to the question: "Is blindness a disadvantage Y/N?" 98% of respondents indicated "Y" in response (Pierson et al., 2024, pp. 14, 15).

Furthermore, belief inconsistent views are subjected to greater scrutiny (Lord et al., 1979) particularly on issues of practical importance. Rebecca Brown observes correctly that unorthodox "positions on contentious [practical/ethical] topics attract far more scrutiny than abstract philosophical contributions to niche subjects. This means that, in effect, the former are required to be more rigorous than the latter" (Brown, 2019). The defender of heterodox views must foresee and head off more potential misappropriations, petty criticisms, misinterpretations, and misunderstandings. Add to this picture the fact that bioethics is also a difficult discipline. The philosophical literature around abortion and euthanasia is analytically rich, complex, sometimes arcane but never vapid. The outcome of our theorizing has consequences. Because of these facts, a brief comment is warranted to describe my zetetic outlook.

In his characteristically undulating and lucid style Brand Blanshard says that the rational temper, "means being realistic, impartial, just; seeing things as they are rather than as fears or desires would tempt one to see them" (Blanshard, 1944, 346). Whereas our desires want to conform the world to our intentions; our inquiry should conform our beliefs to the world. "To be reasonable either in thought or in act requires bowing to an authority beyond ourselves, conceding that there is a truth and a right that we cannot make or unmake, to which our caprices must defer" (Blanshard, 1944, 347). Inquiry aims for a mind-to-world fit. Understood as such Blanshard poignantly concludes, "the best thinking is the least free" (Blanshard, 1944, 351).

All my interlocutors in this project are ones who I think exude the rational temper. I've tried to exude the same here in moving the discourse on abortion and euthanasia "down" to the justifications for such positions. What I hope to reveal are reasons for being much more diffident in one's commitment to (pk).

The present project argues that a permissive stance on abortion and euthanasia does not enjoy considerable justification; the confidence with which bioethicists hold such views is misplaced. The principal arguments for both positions suffer from serious de jure defects (defined below). Just how defective such arguments are might be a persisting point of disagreement, but that there are defects I hope to show convincingly. I have no pretensions that the present project will convince *everyone*, but this work will not be a banal read.

1.3　Nomenclature and Overview

Much of what follows will be using language exported from epistemological discussions, which may be foreign to those interested in bioethics. Some explication of this nomenclature is required.

Imagine the following dialogue in the context of two physicians discussing a patient diagnosed as being in a minimally conscious state[2]:

Doctor A: The patient I am seeing on 2D is now in the minimally conscious state (MCS).

Doctor B: What is your treatment plan?

Dr. A: Well, clearly, I'm going to withhold tube feeding from the patient. I'll have the social worker call the family members to have them come in. The poor patient. She suffers a traumatic brain injury, survives a brief cardiac arrest, and now only progresses to MCS.

Dr. B: Why would you withhold tube feeding from the patient?

---

[2] The minimally conscious state has a technical definition meaning "a condition of severely altered consciousness in which minimal but definite behavioral evidence of self or environmental awareness is demonstrated" (Giacino et al., 2002, 350–51).

Dr. A: Can't you see? The patient has zero quality of life. Nourishing the patient is worthless. She cannot experience anything or at least not anything meaningful.

Dr. B: I see that the patient's quality of life is low. But does having a lower quality of life justify withdrawing nourishment from the person? The patient can demonstrate awareness of her environment. And she might get better over time, it's only been ten days since the admission, right? Furthermore, tube feeding is easy to administer. Does the patient have home care options?

Dr. A: Can't you see? It would be worthless to feed the patient. She's in the minimally conscious state, she can't have meaningful experiences anymore. I can't imagine how much she is suffering right now. It would be best to put her out of her misery.

The disagreement here between these two doctors comes down to a basic perception of value. One doctor sees life in a minimally conscious state as meaningless, and this is supposedly a reason to withhold tube feeding. The other may or may not see being in MCS as meaningless, but what Dr. B certainly does not see is that a lower quality of life *justifies* withholding an easily administered means of sustaining life. The dialogue, not necessarily fictitious, illustrates several epistemological distinctions.

The fictitious dialogue is a starting point from which to understand several epistemic concepts. I begin first with epistemic method. Both doctors are relying on their intuitions of what is valuable. We all have to start there. Bernard Williams observes that ethical theories "have to start from somewhere ... And the only starting point left [is] ethical experience itself" (Williams, 1985, 93). By ethical experience Williams clearly means ethical intuition. "Those initial ethical beliefs are often called in current philosophy *intuitions* ..." (Williams, 1985, 93). But we want our moral judgments on this or that issue to cohere. If Doctor A thinks that being minimally conscious is meaningless, she would have to say the same thing on other cases of cognitive impairment unless there is a principled reason for a difference. The process here is called reflective equilibrium (RE). It is called 'reflective' because one begins by reflecting on cases and principles. It is called equilibrium because one aims to have the results of such reflection cohere.

The basic idea with (RE) is that it is a project of justifying ethical beliefs that "ideally involves the attempt to bring our most confident ethical judgments, our ethical principles, and our background social, psychological, and philosophical theories into a state of harmony or equilibrium" (Arras, 2006, 47). Reflective equilibrium is a method according to which, we start by collating our most considered intuitions (for example, intentionally killing innocent persons is wrong, respecting a patient's reasonable wish is obligatory, and so on). We construct moral principles based on these ground-level intuitions. We then cast our doxastic net wider and wider, ensuring that our intuitions and principles cohere with one another. Arras comments,

> We thus zip back and forth, nipping an intuitive judgment here, tucking a principle there, building up or reformulating a theory in the background, until all the disparate elements of our moral assessments are brought into a more or less steady state of harmonious equilibrium (Arras, 2006, 47).

Ensuring that one's judgments cohere involves what Haslett (1987) refers to as making adjustment decisions. Suppose you initially think that both MCS patients may have their feedings withheld but you also subscribe to the principle that clinicians should care for their patients whether they are conscious or not. Without any "nipping and tucking" this principle may suggest feeding. You have an adjustment decision to make. How do you make it? Haslett notes correctly that RE does not tell you how, you rely on your intuitions throughout. Certainly, there are problems here (Depaul, 1993) but RE is a dominant view in moral epistemology and is accepted by my interlocutors.[3]

An epistemic method yields beliefs. Beliefs can be divided into basic beliefs and non-basic beliefs (Audi, 2001, 32 ff.). Non-basic beliefs are justified on the basis of other beliefs. Basic beliefs can either be justified or not, but not by other beliefs. If they are justified, they are justified because of their relation to a non-doxastic state such as an experience or intuition. Epistemologists typically assume that doxastic states (i.e., beliefs and other contiguous states) have propositional content, by which is meant that having a belief requires having facility with concepts. The belief that it is sunny outside entails that one understands the concept of seeing, and the sun. Non-doxastic states are those that do not have conceptual or propositional content. States of affairs such as it being sunny outside do not have conceptual content (though Hegelian idealists might disagree). When I see the sun, I do not see a *concept*. I see an object, the sun; and when I'm not looking directly at it, I apprehend the state of affair of it being sunny.

Typically, a basic belief is considered justified if the faculty that produced it is reliable or functioning properly (Alston, 1991; Audi, 2001; Goldman, 1986; Plantinga, 1993).[4] Whether the experience or intuition is veridical, e.g., produced by a reliable faculty, will dictate whether the non-doxastic state in question can justify the basic belief. Perceptual beliefs based on visual sensing, at medium distances, in good lighting are examples of justified basic beliefs since they are justified by perceptual experiences which are not other beliefs— assuming that our perceptual apparatus is reliable in such conditions. My belief that the piano I see has fewer than

---

[3] Although I am presupposing RE, some might not like this concession even for pragmatic reasons. But I think that reticence is mistaken. RE is a method of moral inquiry according to which the inquirer relies on her intuitions for every "nipping" and "tucking." Some might be tempted to infer from this that RE presupposes that there are no moral absolutes. If there were, the method should accommodate some external correction on an inquirer's network of moral commitments. Such an inference, however, understands RE as a guide to truth and not a method of moral justification. RE, properly understood, is the latter. Even theorists who endorse moral absolutes, of which I am one, appeal to their readers' intuitions, and provide test cases to motivate their commitments. We 'nip and tuck' to justify our moral outlook.

[4] Debate concerns whether an agent must have "access" to the reliability of one's faculties in order to be justified in one's beliefs. No one I know of disagrees that reliability, if satisfied, is a necessary condition on knowledge.

88 keys is justified, however, on another belief, namely, that I am at a classical period concert purporting to use period instruments (Audi, 2001).[5]

Both types of beliefs have "positive epistemic status" only if they are produced by a reliable faculty, for example, a reliable perceptual faculty in the case of basic perceptual beliefs, and one's inferential faculty in the case of non-basic beliefs. Concerns about the reliability of one's faculties can be extended to apply to one's inquiry, i.e., the *use* she makes of her faculties. One can be diligent, careful, and fair in her intellectual pursuits, or close-minded, careless, and hasty. Conducting one's inquiry in intellectually vicious or virtuous ways are relevant for assessing the reliability of one's inquiry and likewise the epistemic status of the beliefs resulting from that inquiry. So, both one's faculties *and* inquiry are subject to the same question: are they reliable? Because both can be considered information *input* and belief *output* mechanisms, we can ask of both whether they are reliable.

Consider again the fictitious dialogue above. Dr. B does not see the meaninglessness or, does not see the inference from 'meaninglessness in MCS' to 'withholding nourishment from the MCS patient is permissible.' It could be that our imaginations of what counts as a meaningful life are constrained by our experience as fully conscious individuals. Since we have not experienced minimal consciousness, we are probably not in the best position to judge what life would be like in such a state. How, then, should Dr. A respond to the disagreement she encounters with Dr. B? Both are putative peers, they have cared for these patients before, they know the medically relevant details, and they are both trying to be virtuous physicians.

The disagreement bottoms out into a basic belief. Dr. A cannot justify her belief beyond claiming a basic perception of value or absence thereof viz., "can't you see?" On what basis can Dr. A discharge the epistemic effects of Dr. B's disagreement? Since Dr. A's (and B's) judgment is a basic belief, it is justified only if the epistemic faculty producing the belief is reliable. But the epistemic effects of disagreement challenge precisely that judgment, namely, the judgment that one's faculty in producing the disputed belief is reliable. Dr. A could collate her other moral beliefs she thinks are true to form a track record argument for her reliability. But that is obviously circular in that she must assume that in forming her other 'true' beliefs that her faculty was reliable. She must assume that those other beliefs are true, and that requires assuming that they too, are produced by a reliable moral faculty. So, "the fact that you disagree is a reason to think that you and [a putative peer] were out of your depths or that one of you suffered from a performance error" (Littlejohn, 2013, 170). Since you cannot provide a non-circular justification for why *your* faculties and not your interlocutor's are functioning reliably in the dispute, a disagreement is undischarged evidence for a performance error.

Peer disagreement supplies what epistemologists call a defeater to one's belief typically one's belief that one's processing is entirely veridical. Dr. B questions two different aspects of Dr. A's judgment. On one level, Dr. B is questioning the

---

[5] Contemporary pianos have 88 keys the addition of which occurred in the late 1880s. The classical period ended circa 1820.

reliability of Dr. A's moral perception. On another level, Dr. B is questioning the first-order justification for withdrawing tube feeding. Even if it is granted that the patient has a meaningless life, does it necessarily follow that she should not receive nourishment? Part of the skepticism here may be a function of our skepticism in judging meaningfulness, particularly for other people, and our inability to apprehend the conscious states of others.

Regarding these differences, we can distinguish between first-order defeaters and higher order defeaters. First-order defeaters are propositions that if believed by the agent either conflict with the target belief, or conflict with the reasons for the target belief. These are referred to as rebutting and undercutting defeaters respectively (Pollock & Cruz, 1999). Let D be a defeater, p the target belief and r the reasons for p. A rebutting defeater D is inconsistent with the truth of p. I ride by a field on my bicycle and see what look like a sheep. I form the belief that a sheep is in the field. I encounter you, the owner of the property, and you report that there are no sheep in several miles from here, but that you have a dog that looks like a sheep. Your report, if true, is inconsistent with my belief that there is a sheep in the field (Plantinga, 2000).

An undercutting defeater D challenges the support relation between r and p. I visit your widget factory that is pumping out widgets on an assembly line (Bergmann, 2004, 426). I see that they are red. You, the factory foreman, notify me that the widgets are irradiated by red incandescent light to better see any imperfections. This information entails that *they would look red even if they were not.* Notice, this defeater does not defeat my belief that they look red, they defeat the support relation between 'looking red' and 'being red'. They could still be red, I lose my justification, however, for believing that they are red solely on how they look.[6]

Higher order defeaters are propositions that if believed, undercut one's reasons for thinking that her faculty or inquiry is reliable in producing the first-order beliefs. Higher-order defeaters can only come by way of the undercutting variety. The reason is that higher order rebutting defeaters are self-defeating. Let R refer to the belief that my faculties are reliable. Since my beliefs are generated by a faculty, including my belief in the defeater, if the defeater D rebuts belief in R directly, then I have no reason for believing D either. In fact, I have reason for not believing D if I believe that it is produced by an unreliable faculty. So, higher order rebutting defeaters are self-defeating; if I believe them, I have reason for not believing them.

---

[6] Disagreement on the notion of undercutting defeaters concerns just how they in fact *defeat* one's justification (McGrath, 2021; Sturgeon, 2014 and Tiozzo, 2024). For example, to take the foreman's testimony as a defeater I must believe also that my justification for believing that the widgets are red is because they look red. If I believed they are red from some other source, the foreman's comment is inert. The discussion that follows is unaffected by this disagreement since adding this meta-level belief about one's justification for p does not affect the analysis of *that* justification.

1.4 Outline of the Project

The nomenclature of de jure and de facto and the other concepts just discussed may be used throughout the work. They give the project a conceptual skeleton into which my chief concerns serve as the flesh.

This project argues that several orthodoxies present in bioethics are unjustified, unfounded, or viciously circular. Specifically, the chapters that follow consider whether the judgments 'abortion is permissible,' and 'euthanasia is permissible' and their denials suffer from any de jure defects, and if so, how serious are these defects. Since abortion and euthanasia are arguably cases of killing, I refer to these positions as permissible killing, (pk), and their respective pro-life counterparts as ~(pk). Ultimately, the book defends the following argument in outline:

(A) The arguments for (pk) suffer from serious de jure defects, and the arguments for ~(pk) suffer from trivial or much less serious de jure defects.
(B) If the moral risk in being wrong in acting on (pk) is greater than the risk of being wrong in acting as if ~(pk), and the justification for (pk) suffers from greater de jure defects in relation to the justification for ~(pk), acting as if (pk) should not be done
(C) The moral risk in being wrong in acting on (pk) is greater than the risk in being wrong in acting as if ~(pk).
(D) Therefore, acting as if (pk) should not be done.

The next chapter gives a taxonomy of de jure defects and explains moral risk in much more detail. Chapters 3, 4, and 6 argue for the first conjunct of premise (A); and Chaps. 5 and 7 argue for the second conjunct of premise (A). Chapter 8 defends (B) and (C).

The only observation to make here regards premise (B). I take premise (B) to be plausible once we understand the notions of moral risk and the asymmetrical de jure defects that exist between (pk) and ~ (pk). In fact, (B) is plausible even if we drop reference to moral risk or de jure defects (but not both). If arguments for a proposition p and its denial ~p are asymmetrical in terms of the de jure defects they incur, one should not act on the proposition that suffers more defects or defects that are more serious. Likewise, if two propositions are equally probable and they fall prey to the same de jure defects, if any, but they bear asymmetrical moral risk, one should act on the less risky proposition. If both conditions are present, the inference noted in (B) becomes that much more plausible. So, my argument argues that (pk) and ~ (pk) are asymmetrical both with respect to moral risk and de jure defects ((pk) looks poorly on both). Therefore, one should not act as if (pk).

The advance this project makes on these matters is that it presents a more dialectically useful or persuasive analysis for my interlocutors. De jure objections can be developed while granting, for the sake of argument, the truth of the claims being considered. I say this is dialectically useful since it lowers the bar of what I wish my interlocutors to concede. My readers may continue to believe that (pk) is true; I provide no arguments here against that. But it is one thing to grant that one's beliefs are false, and another to grant that they have weaker justification than one previously thought. The latter seems easier to accept than revising one's commitments, maybe many commitments. There is an understandable doxastic phenomenology that accompanies coherent sets of belief. Since the latter concession is easier to justify than the former, the present project marks a significant development in the discourse on abortion and euthanasia.

1.5   Historical Examples of de Jure Defects

A very popular argument for the impermissibility of abortion is typically referred to as the future of value (FOV) argument. The starting point for the FOV argument is to ask what is it that makes it wrong to kill someone? What property or feature makes killing wrong? The next question to ask is whether the developing human being has that property or feature. Whatever this property or feature is that makes killing wrong, it should be a property or feature of the victim *at the time that she is a victim*. Marquis states the basic point this way,

> The loss of one's life deprives one of all the experiences, activities, projects, and enjoyments that would otherwise have constituted one's future. Therefore, killing someone is wrong, primarily because the killing inflicts (one of) the greatest possible losses on the victim (Marquis, 1989, 189).

The argument in syllogistic form reads as follows:

1. Having a future of value is the basis for the right not to be killed.
2. Fetuses have a future of value.
3. Therefore, fetuses have the right not to be killed.

Peter McInerney observes, however, that having a *future* with value requires supposing that there is an entity at one time who may continue to exist so as to experience joys or engage in projects at a later time. He states,

> The unexamined premise in the argument is that a fetus *already* has a future-like-ours of which it can be deprived. For the argument to be convincing, it is necessary that a fetus *at its time* "possess" or be related to a future-like-ours in a way that allows the transfer from the wrongness of killing us persons to the wrongness of killing fetuses (McInerney, 1990, 265).[7]

The de jure defect here is that the FOV argument assumes precisely what is in dispute. What is in dispute is whether the fetus is identical to you or me. (Of course, some might grant the identity claim but claim that at the time that we are in the stage of fetal development, we do not have the right to use another person's body for gestation (see Boonin, 2003)). Presupposing such an identity without arguing for it renders the argument inert— it might be a potent argument for interlocutors who might accept the ontological identity between pre-born and post-born human beings

---

[7] See also Brown (2000) for a similar observation and criticism of Marquis (1989).

(DeGrazia, 2005). Only if one grants the identity between preborn and post-born child can the FOV argument be relevant.

Another strategy for arguing for the impermissibility of abortion is to pick up where the FOV argument leaves off. David Boonin considers what he refers to as the essential property argument. The goal of the argument is to argue that the zygote from which you developed is the same thing as you. Just as you are the same human being even though you go through various phases such as infant, toddler, juvenile, adult; so too your identity goes back before birth, namely, zygote-morula-embryo-fetus-neonate. Boonin (2003, 52) represents the argument as follows:

P1: I am the same individual living being as the zygote from which I developed.
P2: I am a person essentially.
P3: If an individual living being has an essential property at one point in time then it has that property at every point in its existence.
C: The zygote from which I developed was a person.

One can readily see how the essential property argument patches up the defect in the FOV argument. The problem, according to Boonin, is that the essential property argument is circular. At issue is P2 (P2' for Boonin's notation). The term 'person' must be understood in a moral sense in order for the essential property argument to function as an argument against abortion. The moral sense of personhood entails that all persons have a right to life, or at least, a right not to be intentionally killed. On the contrary, Boonin observes, there are "many rights that I possess now that I did not possess in earlier stages of my development. So why should we accept P2' so understood?" (Boonin, 2003, 53). Boonin's point is that we can exist without having certain rights. The argument that you and I exist when abortions are performed is not enough to argue that you and I may not be the victims an abortion. Ontology is independent of axiology. He thinks that "the debate over abortion as largely a debate over the moral status of the fetus" (Boonin, 2003, 54) and that includes the question of whether human beings (viz., you or me) have a right to life from the moment they come into existence. Therefore, the essential property argument, Boonin claims, must assume precisely what is in dispute.

There are also arguments for the permissibility of abortion (pk) that commentators have noted suffer from de jure defects. Most famously, Mary Anne Warren, who eventually defends (pk) opens her most popular article with a criticism of two popular arguments for (pk). The first argument asserts that abortion must remain legally available to avoid untoward consequences of restrictive laws, side effects such as deaths due to illegal abortions,[8] poverty and so on. The second argument asserts that

---

[8] Sometimes these considerations are used not to argue for the permissibility of abortion, so much as to argue for the incoherence of the pro-life position. That position has it that innocent human life should not be killed. The pro-life position can easily accommodate the point by coherently objecting to all abortions including "botched" or illegal ones. Therefore, there are no grounds for incoherence.

abortion is an extension of the woman's right to control her own body. Warren responds to these as follows,

> [T]he fact that restricting access to abortion has tragic side effects does not, in itself, show that the restrictions are unjustified, since murder is wrong regardless of the consequences of prohibiting it; and the appeal to the right to control one's body, which is generally construed as a property right, is at best a rather feeble argument for the permissibility of abortion. Mere ownership does not give me the right to kill innocent people whom I find on my property, … (Warren, 1973, 45).

Warren's point is that the arguments in question take for granted that what is killed by an abortion is not a person. That there might be tragic side effects of outlawing abortion is not itself an argument that what is killed is not a person or that such killing is permissible. Compare with the issue of slavery: pre-Garrison defenders of slavery in the United States noted that emancipation would have tragic consequences on the economy and would not be good for the freed slaves since they would have no work, no home, and no resources (Eltis, 1987). We rightly find such defenses tragically bad.

Similar defects affect more recent popular arguments to the effect that abortion is permissible because it functions as self-defense. Permutations of this argument might claim that pregnancy can threaten the mother's life or health, or that compared to abortion, the latter is safer than giving birth. If pregnancy itself can be considered unsafe, it seems to follow that eliminating the threat— i.e. the developing human being— would thereby be permissible.

The defect in this argument is the same as those canvassed by Warren in the arguments she considers, namely, the argument takes for granted that what is killed is not a person. Consider a permutation of Schwarz's stowaway example (Schwarz, 1990).[9] Suppose a child who no fault of his own gets lost and decides to sleep the night on someone's yacht. Before the child wakes up the owner of the yacht takes it out to sea— without checking to see if anyone was on board (suppose it is a big yacht). At sea, the owner discovers that the child is on board. Suppose that no fault of the child, the child's presence on the yacht exposes the owner to a very small risk of morbidity (e.g., suppose the child has the flu) and an even smaller risk of death. It is still not permissible for the owner to throw the child overboard, effectively killing the child. Readers who think otherwise should consider that the child is yours. Do you throw *your* child overboard because she has the flu?[10] Clearly you don't think about killing him. So, the self-defense argument must *presuppose* but not argue for, that what is killed is not a person.

---

[9] I am indebted here to Rachel Crawford's excellent commentary, "Abortion as self-defense," at https://www.youtube.com/watch?v=B2TakKSUawA&t=1277s. (Accessed, 5/23/2024).

[10] Intuitions can differ on the case for two different reasons. In the text, I focused on the child being innocent and that alone is reason for not throwing him overboard no matter the risks. Others might think that the risks matter. For those readers, consider driving according to which one is leaving one vulnerable to the stupidity of other drivers. That still would not justify killing other drivers even though the risk of morbidity is elevated when one chooses to drive.

Part of the attraction to self-defense arguments is that pregnancy is defined as an attack. This interpretation is especially appealing if the pregnancy is unintended, or the child is not wanted (those can be two different things as with sex selective abortion). The justification for thinking that it is an attack, however, must be that it is against the will of the mother to be pregnant. Pregnancies, after all, are a sign of the reproductive system itself functioning well. Wanted pregnancies are not for that matter viewed as attacks. The implicit premise in a self-defense argument is that if someone's existence frustrates my will, I may permissibly end that person's existence. Stated in this way one can see its underlying assumption, namely, the pre-born child has no independent value or worth. Of course, proponents of such arguments would resist such an understanding; they would no doubt take issue with the reference to person. But that would just admit that the self-defense argument presupposes but does not argue for the claim that what is killed is not a person. In Chap. 3 I explore arguments for (pk) supposing that what is killed is a person.

These historical cases illustrate a common feature: the project of identifying de jure defects aims to plumb background assumptions and asks whether such assumptions would be viewed as plausible vis-à-vis one's target audience. The term 'audience' here should be understood as a constellation of mutually supporting doxastic commitments, not necessarily agents except in so far as such belief commitments require an agent with cognitive faculties to have them.

But there is a difference between historical iterations of identifying de jure defects and the present project. Consider Boonin's complaint that the essential property argument begs the question at stake in the abortion issue. In one sense this is correct: arguing that you and I exist is not itself an argument that you and I ought not to be killed. But arguing that that omission is itself a de jure defect is itself a claim that presupposes a network of doxastic commitments. That you and I exist is *not enough*, he claims, to ground a claim that it is impermissible to kill us. Notice how easy it is to identify putative de jure defects when one can do so simply by referencing an alternative network of coherent doxastic commitments. Imagine, for example, a permutation of the same strategy vis-à-vis slavery. Suppose someone argues that just because slaves are full human persons does not entail that they cannot be owned. Ontology does not entail axiology, they might say. If this were presented as an argument for a de jure defect, it should be viewed as either risible or tragic.

The present point is not to offer a riposte to Boonin's argument (for that, see Sect. 3.6) but to illustrate that these historical iterations of identifying de jure defects is myopic and tenuous in their *procedure*. Arguing that an argument suffers a de jure defect with exclusive reference to one's own coherent set of commitments is rather easy. As Chap. 2 explains, the procedure in this work is more granular. (Boonin, to his credit, argues based on cases that his interlocutors would find illustrative. For example, in some circumstances one person does not have a right to the use of another person's body (viz., cases of refusing organ donation).) These observations bring us to consider what the present project aims to accomplish.

1.6   Chapter Summaries

Chapter 2 discusses various de jure defects and explains the epistemic role of moral risk. The important point in Chap. 2 is to have in mind a taxonomy of de jure defects. One or more of these defects occurs in the arguments considered in Chaps. 3–7.

Chapter 3 turns to the principle content of the book, namely, the justification for (pk). The first justification for abortion considered is based on the time relative interest account (TRIA) of wrongdoing. The goal of this chapter is to argue that the TRIA justification for abortion rights is viciously circular.

The TRIA is an explanation for why certain states of affairs (SOA) can be the object of rational egoistic concern. What *matters* to S is whatever S can take an interest in, and this will be a function of two factors. The first is the importance to S of realizing (or avoiding) that SOA, and the second is the degree of psychological unity between S now and when S realizes (or avoids) the SOA in the future. Harm, then, is defined as the frustration of S's time relative interest in realizing (or avoiding) an SOA. Accordingly, the TRIA justification for abortion rights relies on at least two controversial claims: (1) An agent harms S *only if* the agent violates S's time-relative interests (TRI); and (2) if S cannot be harmed, S cannot be wronged. On the assumption that preconscious human beings cannot take a time relative interest in future SOA's, including the SOA of continuing to live, killing a preconscious human being does not violate that being's time relative interests. Therefore, killing that being cannot be wrong.

The chapter argues that this argument suffers from several defects. Epistemic circularity ensues, for example, when authors try to defend both (1) and (2). A standard procedure is to appeal to one's moral intuitions on cases involving fetal death. Jeff McMahan claims that "most of us believe that the death of a human fetus is not a terrible tragedy, at least not for the fetus itself" (McMahan, 2002, 78). Notice that even if one shares this intuition, it is the explanation for the intuition that either supports or fails to support the TRIA justification. More to the point, the Fetal Death intuition, as I call it, is meant to support the TRIA of harm. But the TRIA of harm is meant to justify the permissible killing of fetuses (pk). And yet, (pk) entails the Fetal Death intuition.

Chapter 4 discusses the justification for (pk) based on personhood. The focus of this chapter is on claims to the effect that preconscious human beings are not persons in the morally relevant sense. Claims of this sort must argue that the relevant members of a moral community must have an exercisable capacity for consciousness or some other rational ability. It is the argument that one *must* have an *exercisable* capacity that is impaled on one of the defects explained in Chap. 2.

An additional point in this chapter is to argue that such "no-personhood" arguments are not even arguments at all for the permissibility of killing preconscious human beings. More precisely, the typical motivations for psychological views of the person highlight intuitions that do not logically affect our judgments on *when* persons come into being. The latter half of this chapter focuses extensively on brain transplant thought experiments and argues that the intuitions these experiments highlight cannot support the view that preconscious human beings can be killed.

Chapter 5 discusses arguments for ~(pk). This chapter is dialectically relevant to head off claims of the sort that pro-life arguments are *just as* bad as the arguments I analyze in the previous two chapters. Dialectically speaking, the arguments canvassed in Chap. 5 presuppose the success of my arguments in Chaps. 3 and 4. Chapter 5 focuses particular attention on what I refer to as the modal argument against abortion.

Chapter 6 discusses the justification for euthanasia ((pk) now understood to refer to euthanasia). The goal of this chapter is to limn the two principal arguments for (pk): the argument from autonomy, and the argument from compassion. The following points are argued in sequential order. First, I argue that the argument from autonomy depends for its plausibility on the argument from compassion— which focuses on the good of ending one's suffering. Suppose, I argue, it is my autonomous decision to think that a life is not worth living if I suffer from a headache. No one would think that that is a good reason to receive physician-assisted suicide or euthanasia. The good of autonomy, if it is such a significant good, requires supplementation from the argument from compassion. The argument from compassion, however, must pick and choose the cases of suffering for which it might be permissible to kill the sufferer. For example, no one would think that the suffering of a concentration camp victim justifies killing him or her. What this and other cases show is that suffering itself is not a sufficient reason to kill someone. This is one among five de jure defects identified in the autonomy or compassion arguments.

Chapter 7 discusses arguments for ~(pk) specified to euthanasia. The dialectical goal of this chapter is the same as Chap. 5 with the obvious difference of addressing arguments against euthanasia.

Chapter 8 discusses the argument from moral risk. One might think that because my argument focuses on de jure defects and not de facto ones, it is dialectically inert. One might still think that the best position all around is still (pk). This chapter argues that such a position cannot be sustained in light of the moral risks in being wrong in judgments involving the *permission* to kill. The focus of this chapter is on authors who have previously addressed the issues of moral risk but say that the risk is entirely on the pro-life argument. Here too, what is novel in this chapter is a closer line-by-line analysis of these authors. Though more tedious, this procedure affords a more dialectically useful exchange where my interlocutors enjoy a direct response to their specific assumptions.

1.7 An Analogy

Think of the present project along the following analogy. Suppose you are a sailor who is undergoing training in a submarine (Otte, MS). To keep the training authentic, you're not told whether the training takes place in a real submarine versus a simulator. The simulator, we may suppose, looks exactly like a real submarine on the inside, and it responds and feels exactly how a real submarine would respond and feel. Of course, within the simulator, all the instruments are conveying fictitious information, from the radar blips, the speed of the slumbering submarine and other marine craft, the direction in which one is traveling, etc. But you don't know whether you are in the real or simulated submarine, at least not based on what the instruments alone are telling you. You don't know, that is, whether the information they are conveying corresponds to reality. The fact that all the instruments provide coherent information is not enough to infer that they are conveying *veridical* information.

Our moral justifications for certain cases of killing are like these instruments. From inside the 'submarine,' we cannot tell that the 'instruments' are correctly reflecting the moral world and the relative values at stake in various ethical issues. The sailor is, in a sense, in an epistemic cul-de-sac in relation to her judgments about the moral world outside. Because of this, she would not be justified in, for example, pressing all the buttons that launch a nuclear missile. She could be in the real submarine, and the risk in being wrong that she is in the simulator is extraordinary. So too for cases of killing. My project begins with this suspicion; the suspicion that one's moral faculties are processing moral reality correctly. The implications of having beliefs that suffer de jure defects are reasons for thinking that one's moral instruments do not track reality. One's network of beliefs *may* be coherent, but they may not function as justifications (see Chap. 2). And when a moral judgment involves a significant risk in being wrong, the agent would be practically rational in not acting on it (see Chap. 8).

Chapter 2
De Jure Defects

2.1 Introduction

Artificial intelligence (AI) systems have been used to assist human decision-making in management, engineering, and medicine. Consider an AI system whose outputs are moral judgments. Suppose the system takes as input descriptions of human behavior, and its output is a moral judgment about that behavior. Under what conditions would I trust the output of such a system? Presumably, the question of trust does not come up if its outputs concur with mine (Whitby, 2008). So, let's understand the context of the question as follows: under what conditions would I trust it if either I have not thought about the issue, or its outputs are discordant with mine?

An initial thought in response is that one would need to know *how* the AI system determined its moral judgment. What was its procedure or algorithm? In what way does it sense or process values or goods? How does it determine the weight or importance of competing values in ethical dilemmas? A common theme in these questions is that they all aim to assess how the AI system is processing moral information (Liu et al., 2022). What is its justification and is its moral faculty or way of processing the relevant information reliable?[1]

---

[1] Objections to the idea that we have a moral faculty are unfounded once we understand what a faculty is in this context. Properly understood, a cognitive faculty is an input-output mechanism. Alvin Goldman states that by a process he means "a functional operation or procedure, i.e., something that generates a mapping from certain states—'inputs'—into other states—'outputs'. The outputs in the present case are states of believing this or that proposition at a given moment (Goldman, 1986, 11). When the outputs are moral judgments, judgments of the kind where the agent is detecting values or goods, such a faculty must process or take as its input values or goods. (Shifting the discourse to focus on moral reasons simply passes the buck and fails to account for our moral experience (Brewer, 2009)). If we make *moral* judgments, we must have a mechanism that processes *moral* information. Non-cognitivists can consistently deny this, but they should also deny that the moral judgments made in this work have any truth value, and that includes metalevel judgments that a putative justification supports or determines a moral claim.

S. Napier, *Justified Killing*, https://doi.org/10.1007/978-3-032-14946-6_2

Now imagine that that system of generating moral judgments is your own moral faculty not some AI system out there, so to speak. Simply moving the question from the third person to the first person doesn't eliminate the importance of the questions canvassed in the previous paragraph. How is it that *my* moral faculty is processing moral information? How am *I* determining the relative weight of competing values in ethical dilemmas? These questions need to be answered before answering the overarching question, under what conditions can I trust it? Presumably, trusting my moral faculty *simply* because it is mine is not a good reason.

This chapter explores the various sources in which we might acquire epistemic diffidence about our moral judgments. The reference to an AI system serves as a helpful heuristic for understanding the issues at stake, namely, how can I know that my processing of moral data is reliable. Call this the reliability question and it captures the questions raised when reflecting on Otte's submarine example. From this broader reliability question, de jure defects descend, if you will, to specific claims about an argument's pattern of justification.

2.2 De Jure Considerations: A First Approximation

Research ethicists look back with opprobrium on the numerous ethical abuses perpetrated by individual researchers and governments. To take one example consider the Willowbrook experiments running from 1956 to 1971. The goal of the study was to test a vaccine for hepatitis. The subjects were mentally disabled children in residence at the Willowbrook state institution for mentally challenged children. The subjects themselves could not consent, and the parents were neither fully informed nor was their consent fully voluntary. Even so, the researchers deliberately inoculated the subjects with hepatitis and then tested a gamma globulin to see if it had a protective effect against hepatitis infection. Subjects were initially infected by feeding them the feces of other children who were infected. Later inoculations were more refined (Coleman et al., 2005).

As it turned out, the researchers were correct that gamma globulin does offer a protective effect against hepatitis, their judgment was true. But were they justified in doing the research in the fashion that they did it? What could justify deliberately inoculating nonconsenting subjects with a potentially serious disease? Even if they were correct in their therapeutic predictions, what *justifies* their actions?

This instance of research ethics abuse illustrates the distinction between claims that are true and having requisite justification to act on those claims. The present book asks what is the justification for the various claims under consideration, not so much whether they are true. For a parallel, consider what many academic philosophers say about theistic belief. Many of these philosophers claim that theistic belief, and Christian belief in particular, is false. But they also say things suggesting that it is irrational to believe that there is a divine being or other religious claims. Alvin Plantinga explains,

> De jure objections, by contrast, while perhaps more widely urged than their de facto coun-
> terparts, are also much less straightforward. The conclusion of such an objection will be
> that there is something wrong with Christian belief—something other than falsehood—or
> else something wrong with the Christian believer: it or she is unjustified, or irrational, or
> rationally unacceptable, in some way wanting (Plantinga, 2000, ix–x).

So, there are two ways to object to a belief: object to its truth, and object to its justification or rationality.

2.3   De Jure Defects: A Second Approximation and Taxonomy

There are two tasks in this project. The first is addressed in this section; to explain
the kinds of de jure defects and their effects. As to their effects, Michael Depaul's
smelting analogy (2004), is helpful: de jure defects can turn what *an agent thinks* is
epistemic gold into what *is really* epistemic lead. The second task is to argue that
this or that argument suffers from a de jure defect. The latter task constitutes the
labor of Chaps. 3–7.

For simplicity, focus attention on a justification J for a proposition P.[2] For any
justification J, there is an inferential claim to the effect that it supports, entails, or is
otherwise epistemically related to P. Call this the inference. We can consider the
justification, the inference, and the proposition being justified (abbreviated as J,
inference, and P respectively). Call one's processing of or deliberations about J,
inference, and P, one's inquiry. Lastly, assume that one's inquiry requires making
use of one's cognitive faculties.

<u>Diagram: de jure defects taxonomy</u>

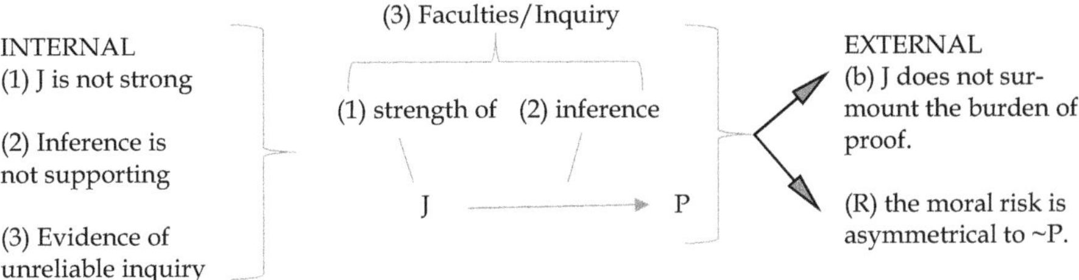

With this division in mind, there are a total of five different categories of de jure
defects under two broad headings, internal and external. By internal I mean internal
to the agent's processing of J, inferring from J to P, or forming the belief that
P. Details on these divisions will follow, but understanding these divisions requires
a word or two about the term justification.

---

[2] At this point, it is not important to distinguish between reasons and evidence.

The term justification in ordinary usage, and as I am using it here, is usually a quantity term according to which one can have strong or weak justification; or gain or lose it. I say it is usually a quantity term since sometimes we use it to refer to an argument. Arguments themselves are clearly not more or less. I mention this usage because below I note the de jure defect of circularity which is a defect of arguments, not of propositions or evidence. The context will make it clear how the term is being used.

Justification for epistemologists means something like the term evidence for scientists. Furthermore, whereas obtaining knowledge is an inquiry stopper (Melchior, 2019), having justification is not. If an agent acquires a reason for her beliefs, that by no means justifies stopping one's inquiry. Whereas if someone knows that P, why inquire further? Finally, as for the term itself various metaphors come to mind. Being justified in believing P suggests having "support" or "grounds" for P (Alston, 2005). Other usage requires other metaphors. I might be justified in defending myself against an unjust attack. Being 'justified' here means that the circumstances *remove an obstacle* that would otherwise make my defensive actions wrong. These examples suggest that the term is primitive and its meaning is understood in the context of its use.[3]

In saying this, I am setting to one side the question of what it is to be justified (Alston, 1985, 57)? What does the concept *mean*? Answers abound. There is the deontological notion of justification according to which an agent has a justified belief B if and only if the agent has not violated any epistemic obligations or duties in forming the belief that B. There are virtue accounts of justification according to which an agent has a justified belief B if and only if the agent has formed B in a virtuous way (Zagzebski, 1996, 233 ff.). And there are positions such as coherentism, foundationalism, internalism, externalism, evidentialism, and so on that advertise as being accounts of justification but look nothing like deontological or virtue accounts. What can one do with such a pastiche of viewpoints? It is enough here to make two observations.

First, the diversity of answers these theories give suggests a diversity of questions. For instance, deontological and virtue accounts of justification seem more focused on explaining why having a justified belief is *good*. What these ethical theories do for action, they can do for beliefs—i.e., explain their epistemic goodness.

---

[3] This observation directly contradicts Alston (1985) and many others, so a brief comment is warranted. Our intuitions on whether an agent is justified in this or that circumstance seem much more secure than abstract definitions of justification with numerous embedded conditions. A risible example of this is Swain's definition of epistemic justification (1981, 133–134) which runs a page long and includes three main conditions, three sub-criteria each of which includes two further conditions. Do I know that my belief that $2 + 2 = 4$ is justified only after I understand such a definition? Clearly not. And I am more justified in believing $2 + 2 = 4$ than I am in believing Swain's or any other definition of justification. What is more, when any of us use the term justification we clearly do not *mean* "S's believing that h on the basis of R is a reliable indication that h at t iff: there is some maximal set of relevant characteristics, C, such that ..." (Swain, 1981, 133).

Internalism versus externalism are answers to the question 'under what conditions is an *agent* justified in believing B'—as opposed to what a justified *belief* is. Internalists think that an agent must have "access" to the justification for one's beliefs; externalists think that so long as an agent's epistemic faculties are reliable, the agent is justified. (Another interpretation is that internalists are limning when an agent is justified, and externalists are liming when an agent knows (see Bonjour, 2009, Sect. 7)). This is not the place to apportion these discussions to their proper destinations.

Second, what is important to understand is that a defect of justification requires understanding the role or function of justification. A defect suggests an imperfection of something, or a falling away from a norm. A paradigm instance is required to identify a defect. Most will agree that being justified pertains to either beliefs or agents; it is an epistemically good thing to have; and justification comes in degrees (Alston, 1985). Most will also agree that being justified *functions* to dispel doubts, respond to challenges, or to answer questions. Wellman states the point this way, "[w]hatever justification may be it must include meeting questions and doubts" (Wellman, 1971, 116). What is necessary in what follows is to understand what being justified *does*. The work of justifying one's beliefs aims to make changes to one's noetic commitments—either making one more (or less) certain in a belief B or changing B altogether. De jure defects are understood in relation to this functional role of justification. So, a weak justification will be one that gives an agent no reason to modify her noetic commitments.

2.3.1   Internal Justification Defects

The first iteration of a de jure defect is an internal-J defect (hereafter justification-defect). This occurs when the justification J for P is *weak* and it can be weak in one of the following four ways: (i) There exists a similar set of justification(s) but a different conclusion follows. This is explained in further detail below, but the basic idea is that some justifications for a moral claim P appeal to the same values shared by those who believe ~P. This suggests that having J does not warrant a shift in one's epistemic commitments. (ii) There exists a separate set of justifications that are equally intuitive but they support ~P. Here too, having J does not warrant a shift in one's commitments because there are equally intuitive reasons for ~P. (iii) The previous two conditions are mirror images of each other. The third condition departs in kind in that J may avoid both (i) and (ii) but still not warrant a shift in one's epistemic commitments. This can happen if J only includes a subset of *relevant* reasons, namely, reasons that do not meet some external standard such as a burden of proof in legal contexts. Finally, when we say that a justification is weak, we can use that phrase to mean simply that (iv) the justifications are implausible. The following diagram represents these divisions.

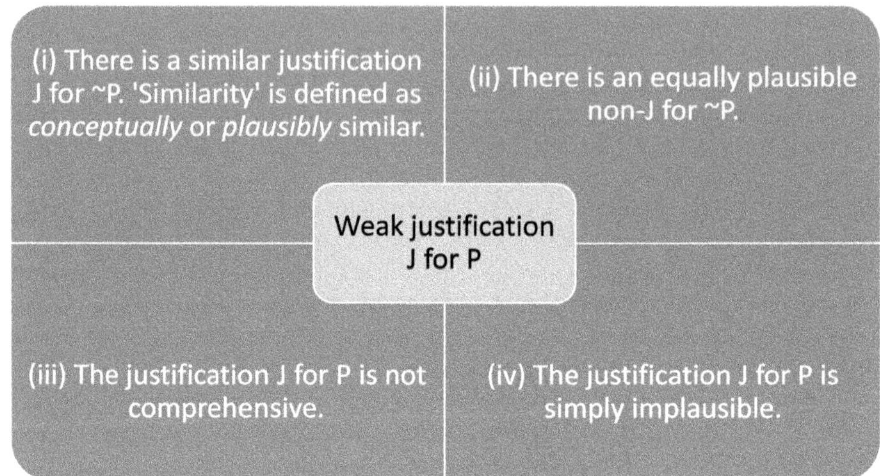

These are the distinctions for the notion of weak justification that capture a good chunk of ordinary usage. Some examples, particularly of (i) and (ii) are required to see why these are defects.

The famous legal case *Board of Regents for the UCLA v. Tarasoff* (1976) illustrates aspects of condition (i). A psychiatrist employed by the UCLA medical school was seeing a patient named Poddar. Poddar was receiving psychotherapy for several issues including aggression and anger. He had reported to his psychiatrist a desire to kill his ex-girlfriend Ms. Tarasoff. The psychiatrist did not warn the family or the law enforcement authorities of Poddar's intentions. Unfortunately, Poddar acted on his intentions and killed his ex-girlfriend. The family of the girlfriend sued the Board of Regents for a failure to warn.

The California Supreme Court ruled 2-1 in favor of a duty to warn. The two judges in favor argued that a duty to warn policy in the setting of psychotherapy ensures the health and welfare of relevant stakeholders. The lives of those outside the therapeutic relationship matter. The value of their lives is a more important value than a strict protection of confidentiality.

What is interesting about this justification is that the dissenting Judge argued to a different conclusion based on the same values—there was a conceptually similar justification. If therapists had a duty to warn and their patients knew about this, the patients might be less willing to divulge violent thoughts. But if violent thoughts are not shared in therapeutic encounters, they cannot be treated. If they are not treated, the lives of those outside the therapeutic relationship are threatened. *Because those lives matter*, Judge Clark argued, confidentiality should be strictly protected.

Here is an example where the justification for a duty to warn could be considered weak in so far as it is largely compatible with similar reasons that support no such duty. Both appeal to the same ethical values, namely, protecting the lives of those outside the clinical encounter.[4]

Of course, one needs to argue that there is a defect here. The basic idea with condition (i) is that if there exists a set of justifications J1 and a very similar set J2 and each justifies different conclusions, condition (i) may be met. *Arguing* that condition (i) is met may take different forms. One might argue that the features that J1 and J2 do not share, if any, are irrelevant or not in dispute. Another option is if J1 and J2 overlap significantly enough, then arguing that (i) is met requires arguing how an opposite conclusion coherently follows.

Apropos of the discussion that follows (see Sect. 4.4.1), condition (i) is met also in some cases where a thought experiment is used as an argument for P. Michael Bishop (1999) points out that if the same thought experiment can be used to support two different conclusions, the thought experiment cannot function as an argument for either conclusion. In such a scenario, disputants may share the belief that thought experiments can be philosophically illuminating, they can agree that they have the same "intuitions" when considering the thoughts experiment, but they can disagree either about the explanation for their shared intuitions, or about the purported results of the experiment. Importantly, the thought experiment itself does not dictate which explanation is best, nor may it dictate, as with real experiments, its results.

Condition (ii) should also be familiar to many readers. One way of interpreting the discussion on conscientious objection is that disputants recognize the value of autonomy, but one side emphasizes patient autonomy, the other clinician autonomy. (There is, of course, an asymmetry. If a doctor refuses to perform euthanasia on a patient who requests it, the patient's conscience is not *violated*. Whereas the clinician's conscience would be violated if he were forced to perform euthanasia. This asymmetry might not make a moral difference, but it is relevant to be precise about whose will is violated.) This is not to say this condition is a version of condition (i). The disputants on conscience rights endorse different sets of values (patient autonomy versus clinician autonomy). What gives the discussion a perennial quality to it is that it is difficult to adjudicate which value should trump the other (see Napier, n.d, for a way to obviate this stalemate).

One might wonder if all that is required for a justification to count as weak is that there exists nonidentical justification for an opposite conclusion, then what's the defect? Numerous arguments in ethical discourse are shot through with such epistemic features but we do not think that they are all weak. In response, what makes moral discourse interesting is that many counter-arguments exhume previously

---

[4]Those who think that a duty to warn is obvious and that the dissenting opinion is wrong should consider the deleterious effects of such a policy. See Wipond (2024). https://www.madinamerica.com/2024/06/dramatic-rise-in-police-interventions-on-988-callers/?mc_cid=ac64d05edb&mc_eid=1b5e96dfbe. Wipond documents that alerting authorities can invite police brutality and needless psychiatric incarceration. Furthermore, Soulier et al. (2010) reviewed 70 appellate cases from 1985 to 2006 that involved claims against a therapist for failing to warn. 66% (or 46 cases) ruled in favor of the therapist. Clark's reasoning clearly circumscribed the strength and scope of the majority opinion in *Tarasoff*.

unrecognized aspects of our reasoning. The merit of a journal article, after all, is partly a function of its insight on previously hidden or subtle features of an argument that challenge its justification. So, it is not at all counter-intuitive to say that many arguments for various ethical positions are in fact weak. In saying that they may be weak, however, I am not saying that arguments for de jure defects are *convincing*. To argue that an argument satisfies a condition (ii) defect requires arguing that the there is in fact *equally plausible* justifications for ~P. If this can be shown, then satisfying condition (ii) is clearly a de jure defect.

One can argue that condition (ii) is met by limning counter arguments or counter evidence. Such arguments come closest to de facto objections, the difference being that de facto defects say that either the justification or the proposition are false. The role of counter arguments in highlighting a de jure defect aim only to show that the justification in question is not itself strong because there are equally plausible alternatives.[5]

Condition (iii) according to which J is not comprehensive is at once a more frequent occurrence, but it does not get attention in the present project. It might occur more frequently at the bedside in clinical settings where arguments for or against withdrawing life-sustaining treatment from a patient focus exclusively on the patient's wishes and not on the benefits (or harms) of the intervention in question. Futility cases (Chwang, 2009), for example, are cases in which the patient's wishes are not of exclusive importance. Discussions on whether a research project on human subjects is ethically permissible can sometimes focus exclusively on whether the subjects are properly informed of the risks. But this is myopic. Risky research on competent adults should also consider the expected harms to the subjects (Napier, 2013; Weijer, 2000; Elliott, 2011, Chap. 1).

So, a justification-defect can occur when two arguments appeal to the same value but yield two different conclusions; there are other justifications non-J that are *equally intuitive* and they support ~P; the justification J is not comprehensive; or J is simply implausible.

2.3.2   Internal Inference Defects

An inference-defect is when the inference from the justification J to P might not be supporting, entailing or determining, for example, P suffers an undercutting defeater (see Sect. 1.3).

A paradigm instance of an inference-defect is some iteration of a non sequitur fallacy. Other terms or metaphors can be used to describe this defect such as that the argument fails to *support, determine,* or *motivate* acceptance of P. Arguments for inference-defects aim to highlight undercutting defeaters. An argument that such an

---

[5] One concrete example of this defect in the present project is how the too-many thinkers problem (Olson, 2002) counter-balances intuitive evidence in favor of psychological accounts of the person (see Sect. 5.4.3 for details).

argument commits this defect must show that P does not follow from J, or that J requires supplemental assumptions that include P or are entailed by P. Some further explanation is required to explain the latter two disjuncts. What is the role of these assumptions?

Helen Longino (1990) discusses the conditions under which some state of affair might count as evidence for a hypothesis or theory. "Any singular state of affairs, such as the level of mercury in a glass tube, of itself, points nowhere" (1990, 40). Her point is that states of affairs do not tell you what they are evidence for, they do not have address labels (Longino, 1990, 40). What is needed are background assumptions. She uses several examples to illustrate the point, one will be enough here. She sees red spots on her daughter's stomach and infers that her daughter has measles. The inference is innocuous enough, but she could have inferred anything from red spots on the stomach such as that the aliens are coming. Red spots on the stomach in and of themselves do not function as evidence *for measles* without background modern medical knowledge. In a helpful analogy she says that the "function of background beliefs is analogous to the function of background conditions in causal interactions. In an atmosphere that contains oxygen if one rubs two dry sticks together a flame…results" (1990, 43). The role that background beliefs play in connecting evidence to hypothesis, she argues, is essential. The epistemic problem is that the background beliefs can be chosen in a self-serving manner; viz. background beliefs are chosen to understand such and such state of affair (SOA) as being evidence for P, but there is no independent reason for choosing those beliefs.

There are two levels at which one may direct an argument that an inference-defect occurs. One may focus on the background assumptions themselves that allow for or permit the inference from J to P. As just indicated, these background assumptions might be chosen in a self-interested manner. To provide a concrete iteration of what this might look like, and to foreshadow a bit, I argue in Chap. 4 that typical interpretations of brain transplant thought experiments are that they provide evidence that personal identity should be understood psychologically as the enduring identity of the same *functional* brain. I argue, however, that interpreting such thought experiments as evidence for a psychological account of the person must import background assumptions. Other interpretations of those same thought experiments are compatible with conceptions of the person understood as being a rational animal, or a rational substance. These latter notions, as will be explained, are compatible with the idea that persons come into existence immediately after successful conception. Therefore, the brain transplant thought experiments do not provide exclusive evidence *for* persons being psychological entities *who do not come into existence until consciousness is realized.*[6]

---

[6] It may be that we are psychological entities in the sense that a human person is the kind of thing that develops an exercisable rational power. In eaglet does not have wings, but it is the kind of thing that develops wings and exercises flight. Persons can be understood as a substance the nature of which develops rational capacities, and who can later exercise them. Importantly, brain transplant thought experiments are logically compatible with persons understood as a substance with a rational nature.

Whereas background assumptions mediate an inference from J to P, the second level of defect can occur in discourse with other views. Defects at this second level might occur when rebutting counter examples to one's views. As is typical in analytical philosophy, a proponent of a position provides justification for P; an opponent to P generates a counter example to either J, or P; and then the original proponent argues that her argument can circumnavigate the counter example.[7] An instance of an inference-defect would be if the proponent of P circumnavigates the counter example by assuming P itself. In fact, my argument in Chap, 3 is that the time relative interest argument for abortion rights recapitulates this very pattern in its response to so-called prenatal-impairment counter-examples (see Chap. 3, Sects. 3.1 and 3.2).

Arguments that an argument suffers a de jure defect of the inquiry variety aim to show that one's inquiry or processing of the morally relevant information is incomplete or potentially faulty. This can be done in two ways. First one can show that (i) one's faculty may not have processed correctly or (ii) one has not considered all morally relevant information. One could also highlight the use one makes of one's faculties, and this I have referred to as one's inquiry. The literature typically refers to either challenge as higher order defeaters or more generally, higher-order evidence (Christensen, 2011; Kelly, 2013, 2008; Lasonen-Aarnio (2014)). Such evidence can be mined from contemporary empirical research on biased processing, or motivated reasoning which is likely ubiquitous in moral discussions (Kunda, 1999). But even if it is not ubiquitous, because such biases and motivations do not announce themselves to the agent who has them, one could unknowingly process information in a biased manner. And the probability of the hypothesis that one might be biased is inscrutable—the probability might be high or low. The fact that one cannot rule out the probability of biased processing is itself a source of diffidence.

A more concrete form of higher-order evidence is peer disagreement as explained in the previous chapter (Sect. 1.4). Consider the issue of peer disagreement in the setting of believing a risky judgment. Richard Rudner noted long ago in the context of nuclear bomb testing the following.

> [O]ur decision regarding the evidence and respecting how strong is "'strong enough'", is going to be a function of the importance, in the typically ethical sense, of making a mistake in accepting or rejecting the hypothesis.... ... How sure we need to be before we accept a hypothesis will depend on how serious a mistake would be (Rudner, 1953, 2).

Suppose you accept the hypothesis that testing this nuclear device would be safe for the environment. Suppose also that you discover that your colleague arrived at a different judgment. Do you lesson your credence? Do you become more diffident in your judgment that it would be safe? Or consider the dialogue from the previous chapter between Doctor A and B. Suppose you are the patient and you do not think life is meaningless even with MCS+ (Graham, 2018). Do you want Doctor A to recalibrate his judgment that you should not receive nourishment?

---

[7] Actual examples of this pattern can be seen in the extended discussion of Gettier-type counterexamples in analytical epistemology (Gettier, 1963).

Higher-order defeaters call into question the reliability of your faculty or inquiry. Peer disagreement exerts skeptical effects that are more direct than other higher-order defeaters like the empirical evidence on biased processing. Faced with a peer who disagrees, I could respond in one of two ways. I could epistemically double-down on my belief and use it as an argument that my putative peer is not really an epistemic peer. Conversely, I could keep the belief that he is a peer and use it to argue that one of us has made a processing error.

If we have the intuition that peer disagreement exerts epistemic effects, a plausible explanation for why is that we want to avoid blatant circularity as with the first response. David Christensen notes that the motivation behind a conciliatory stance to disagreement is "to prevent blatantly question-begging dismissals of the evidence provided by the disagreement of others" (Christensen, 2011, 2). Using my own beliefs to argue that others are not epistemic peers can amount to an egotistical privileging of my own beliefs simply because they are my own. Michael Huemer calls such a position "agent centeredness" and comments that "[a]gent-centeredness seems to call for a kind of epistemological egotism … Each agent seemingly must say, '"my experiences, considered as such, are prima facie better indicators of reality than the experiences of others"'" (Huemer, 2011, 24). On this view, how things seem to other conscientious agents provides me with little to no prima facie justification.

One way to reveal the implausibility of agent-centeredness is to explore what things might look like behind a Rawlsian veil of ignorance. Pasnau exports Rawls' veil of ignorance thought experiment to the effect we are to imagine that you and an interlocutor are disputants to a debate but with the following twist, we are to set aside *who* we are in this debate. We are to set aside which position is ours.

> [L]et us imagine ourselves informed about all the factual circumstances of the situation—what is agreed-upon, what is contested, what the opposing arguments are, what the credences are of the contending parties—but without knowing *who we are* in the dispute. To say that we would not know who we are means… that we would know neither where we stand on the proposition in dispute, nor any other relevant propositions that are contested by the two sides (Pasnau, 2014, 624).

The point of the thought experiment is to motivate the view that one's own beliefs should not get "special weight just because they are my own, or just because they follow from what seems right to me" (Pasnau, 2014, 625). While it is true that rationality demands that I believe what I think is true; by viewing a disagreement from behind the veil, I find that I have no reason for preferring my view *simply because it is my own*. Commonsense norms of rationality permit modulating my moral judgments in relation to the views of others. So, when confronted with someone who I initially think is a peer and she disagrees with me, I suffer a defeater to my belief that I have handled the evidence well. Littlejohn is worth quoting again on this point: "the fact that you disagree is a reason to think that you and Tilda [a putative peer] were out of your depths or that one of you suffered from a performance error" (Littlejohn, 2013, 170). And since you cannot provide a non-circular justification for why your faculties are functioning reliably in the dispute, a disagreement is undischarged evidence for a performance error.

The principal objection to the claim that peer disagreement warrants reducing one's credence in the disputed belief is if the disagreement is characterized as "deep." Deep disagreements are disagreements about the more basic epistemic methods or principles upon which one forms one's beliefs.[8] Though authors support this line of argument (Kappel & Andersen, 2019; Elga, 2007), it is obviously circular. If peer disagreement about first order beliefs warrants reducing one's credence in those beliefs, then disagreement about more basic epistemic methods and principles would warrant reducing one's credence in those methods and principles. Again, peer disagreement supplies higher-order evidence that one's faculty or inquiry has made a performance error. Since we cannot get out of our own faculty, so to speak, we have to trust it; but putative peer disagreement gives one a reason for not trusting it.

Both sources of diffidence (peer disagreement, and biased processing) aim their sites on the reliability of one's processing, and this processing pertains to the acquisition of justification. They call into question the claim or insight that such-and-such justification in fact supports or determines P. Such a de jure defect looks like a de facto one, but the conclusion of the argument that, for example, peer disagreement on P is a de jure defect is that one *lacks justification for trusting* or acquires *justification for distrusting* her faculty that generates belief in P or the belief that J supports P. The skeptical effects of peer disagreement can exert themselves without taking a stand on the truth value of the disagreement (Lackey, 2011).

One may wonder how it is that if I assume reflective equilibrium, as is explained in Chap. 1, I can consistently claim that peer disagreement can function as a defeater. If our intuitions are the start of any moral theorizing, would not that require being agent-centered? In response, there are beliefs we have on the subject matter, and there are beliefs we have on how we are thinking about the subject matter. It is perfectly reasonable to calibrate our thinking on our own inquiry with reference to the

---

[8] Typical examples used to motivate why deep disagreements do not have skeptical effects is by comparing a legitimate epistemic method with one that one's readers would likely find unreliable: e.g., reliance on scientific evidence versus biblical interpretation. But let's substitute different methods to see if this procedure of obviating deep disagreement works as advertised. Suppose someone manifests a deep suspicion of intellectual homogeneity because historical instances of homogeneity were strongly associated with uncritical acceptance of oppressive or discriminatory practices (Alexander, 1949). Or suppose she is suspicious of certain viewpoints because of how they are used. Both suspicions might be directed towards, for lack of better term, a pro-choice view. Clearly such a view is homogenous in academic circles, and so, one would exercise considerable intellectual autonomy in resisting such homogeneity. Roberts and Wood observe that a religious person demonstrates autonomy in the setting of contrary epistemic pressure such as secular culture. "The firmer and more articulate and nuanced she is in her resistance to the secular teachings, the more intellectual autonomy she displays" (Roberts & Wood, 2008, 277). If the homogeneity is widespread, and functions as the default position, it takes more autonomy to critically assess it. One can understand this as a method in so far as an agent conducts her inquiry to exude the intellectual virtue of autonomy or intellectual courage. So understood, this is not a questionable method. But in the case at hand, autonomy directs one to critically assess the widespread acceptance of abortion (or euthanasia). One may choose not to adopt this method, but to disagree with it does not obviate the epistemic effects of peer disagreement.

'findings' of other inquirers. Agent-centeredness, in essence, ignores this calibration.

2.3.3   External Defects: Burden of Proof and Moral Risk

Inquiry-defects are when the inquiry might be faulty, impotent, or incomplete. External defects are twofold: external-burden defects are when the justification for P does not surmount a legitimate burden of proof. External-risk defects are when the justification for P fails to offset the moral risk in being wrong and acting on P (*in relation to* the risks in acting on ~P).

These external de jure defects grant that there is a justification for P that is supporting, but the justification is not strong enough to overcome either the burden of proof, or the epistemic effects of moral risk. The burden of proof or the moral risk (for which there can be overlap) are external to the epistemic relation between one's reasons for P, and P.

A good example of an external-burden defect is the discourse on euthanasia. What might the burden of proof be, and how do interlocutors identify *who* bears this burden? One might frame the issue of euthanasia as asking whether healthcare professionals may permissibly intend to kill their patients who are innocent human persons. It is plausible to presume either (a) that healthcare professionals are not the kind of people who should kill one's patients, or (b) that it is impermissible to intentionally kill an innocent person, period. In favor of the first disjunct, the AMA's code of ethics prohibits any healthcare professional from not only performing but being at all involved in state-sponsored executions. It states that "[a]n individual's opinion on capital punishment is the personal moral decision of the individual. However, as a member of a profession dedicated to preserving life when there is hope of doing so, a physician must not participate in a legally authorized execution" (AMA Code of Ethics, 9.7.3.). Without taking a stance on the moral permissibility of state-sponsored executions, the AMA's Code delineates that as a healthcare professional, it is contrary to the ends of that profession to intentionally kill a person even if such killing might sate retributive justice. One may presume, then, that healthcare professionals are not the kind of agents who may intentionally kill. Such a presumption creates a burden of proof for those who argue against it.

In favor of the second disjunct; there has been a prohibition against killing innocent persons and this prohibition has, historically, not admitted any exceptions. Anyone who argues that there is an exception to this long-standing rule must prove it. The American system of justice presumes innocence, for example, and so those who argue for guilt must prove it. In many cases, they do, illustrating how the burden of proof is external in the sense meant here. The burden of proof sets the goals of the discourse no matter what the content of that discourse is.

Ideally, identifying who has this burden should do so without appeal to specific claims in the debate. Let's return to the example of euthanasia. One could start the discourse with the presumption that it is impermissible always and everywhere to intentionally kill an innocent human person just as one starts legal proceedings with the presumption of innocence. This is not to say that one cannot *argue* in favor of the presumption—viz., healthcare professionals ought not to intentionally kill their patients. And this is also not to say that the burden of proof might shift as the discourse evolves and develops. But in general, interlocutors should set the burden of proof without appealing to specific contents of the debate.

What might violating this criterion look like? One might argue that *there is* a presumption in favor of euthanasia thereby creating a burden of proof for the opponents to euthanasia. One might presume that healthcare professionals should manage someone's suffering to the extent possible; and if it cannot be managed, the sufferer should be eliminated since intractable suffering is unqualifiedly bad, or finally, one should honor a patient's autonomous wishes.

One could argue this way but notice that such an argument uses the specific content of the debate to argue in favor of a particular presumption. The specific content, or claim is that the value of the sufferer's life waxes and wanes depending on the amount and manageability of the suffering. But that claim is the central and contestable claim in the debate on euthanasia. One cannot presume that which is the very object of dispute. Moreover, that a patient's suffering might be intractable certainly does not entail permission to kill that patient. As a practical matter, managing someone's suffering can be done by manipulating the sufferer's *consciousness* of it through, for example, palliative sedation. Such a practice does not require an intention to kill the patient or even an intention that the patient be dead (see Taboada, 2015).

How might (a) and (b) avoid similar reflections? The reason is simple: we assume either (a) or (b) in every context other than discourse on euthanasia—they are not specific to the euthanasia issue. When one arrives, so to speak, at the euthanasia issue, one may justifiably transport (a) or (b) as anchoring the burden of proof. Such assumptions may affect the debate, but they are not specific to the debate. The euthanasia proponent must start the discourse arguing that euthanasia is an *exception* to (a) and (b).

Three points need to be kept in mind to evade the charge of 'stacking the deck' against euthanasia proponents. First, setting the initial burden of proof with reference to the very contestable claims at issue works both ways and can be a recipe for deadlock. Assuming we should avoid deadlock (certainly at the beginning of one's inquiry), the burden of proof should be set without appeal to the very claims at issue. That healthcare professionals should not be the ones who kill persons does not decide whether euthanasia is permissible because that assumption is held constant on the issue of executions. Furthermore, we typically do not alleviate suffering by killing the sufferer (e.g., children living in poverty, POWs, suicidal patients). So, there is no assumption made specific to the euthanasia debate when the presumption is that healthcare professionals may not kill their patients. Second, the burden of proof can shift as the discourse evolves and develops. Setting the burden of proof is not itself an argument that this or that action is permissible. In saying that the burden of proof is on the euthanasia proponent to argue that healthcare professionals can in this one case kill an innocent person does not logically entail that euthanasia is impermissible. Third, burden of proofs can be met. Setting a burden of proof for P does not entail that the justification for P cannot meet that burden. It just means that we should be very cautious about *taking P for granted.*

Let's consider moral risk. In most every context in which the term risk is used, it means there is a probability or likelihood of harm or wrongdoing.[9] The notion of probability should not be understood in strictly statistical terms, though it may include that. It may be any set of reasons for thinking that a harm or wrong might occur. Reasons do not have to be reducible to a statistic.[10] Probability may be understood subjectively (given what the agent believes) or objectively (given the likelihood of harm with reference to the "space of reasons" (McDowell, 1997). Harm or wrongdoing can also be understood subjectively or objectively. Jones is an elderly cancer patient who does not care whether his choice of chemotherapy drug causes infertility. To him, there is no risk of harm at least in that respect. Conversely, a commercial developer should care that his proposed construction of a 120-acre mall on riparian woodland may have devastating ecological impacts.

The notion of risk as applied to our choice of action usually conforms to the following pattern. Consider an action and the morally relevant characteristics of that action (e.g., its consequences, the agent's intentions, circumstances, the object of the action or what the action is and so on). These morally relevant characteristics should be understood as possible considerations—i.e., the space of reasons. They might not be what the specific agent thinks are relevant, but they could be relevant. Suppose that on some subset of morally relevant considerations, the action is wrong or that some harmful outcomes may occur. It is now epistemically probable (given a subset of the space of reasons), that the action involves harm or wrongdoing. Considerations of risk ask how should the agent act in such settings?

The notion of moral risk in the present book is an external constraint on the justification for an action; it applies to what an agent is justified in doing. Considerations of risk apply to actions not beliefs. The reason is that considerations of risk, or uncertainty, or other cognates do not themselves provide *evidence* for or against the belief that the action in question is permissible or not (Horton & Ross, 2025). Of course, probabilistic judgments are admissible evidence in adjudicating the permissibility of an action. For example, the probability of expected harms and benefits for a medical research protocol are constitutive evidence for whether that protocol is permissible. To appreciate the contrast I'm suggesting, consider two very different examples. Typical arguments for God's existence might be some form of the cosmological argument, or ontological argument, etc. Pascal's Wager, however, is not itself an argument for or against God's existence. Pascal's Wager is an argument for how one should handle the evidence one has. The idea of "handling evidence" is difficult to get a handle on, but the intuitive idea can be appreciated by considering whether a person's belief in God would count as justified if it were based *purely* on

---

[9]The term harm in contemporary philosophy typically means a frustration of one's interests or desires. The category of wrongdoing may admit of broader concerns.

[10]See Plantinga (1993) for a discussion of *epistemic* probability.

a Pascal type Wager. Plausibly, the answer is no. Likewise, considerations of risk in this book do not constitute evidence for or against a particular belief.

Prior to the creation of the first large particle supercolliders, some scientists were concerned about several theoretical risks including the creation of mini-black holes, and the creation of new particles since the supercolliders were initiating energy reactions at scales previously untested. These new particles might have energy expenditures or utilization requirements that are unknown but possibly devastating. Many of the theoretical worries involved chain reactions that would destroy the planet (Fantl & McGrath, 2009). So why did they do it? Though the possible outcomes involved significant harm at an apocalyptic scale, there was no known reason for thinking that particles in the collider would behave in such a fashion. The *risk itself* did not give the scientists *reasons* for thinking that the particles would behave as feared. This example introduces a common motif throughout the present discourse. The gravity of harm or wrongdoing if one were wrong about A is not itself a reason against doing A. Rather, moral risk is a *relation* between the justification for A, the justification for not-A, and the gravity of wrongdoing if one were wrong.[11]

So, moral risk is an external constraint on the justifiability of an action. How does it constrain? What does risk do? Consider again Rudner's point above and the exchange between Doctor A and B in the previous chapter. Clearly the risk in getting one's moral judgment wrong should factor in whether we perform the action. But how should it factor? What exactly is risk *doing*?

There are basically four distinct roles that risk might play, though I endorse that only one or more of the first three function on the arguments in this book.

1. Risk might be considered relevant in adjudicating *who* in a dialectical exchange bears the burden of proof. Conceptually speaking, rick and burdens of proof are quite different. But risk can distribute to whom in a debate bears a burden of proof—and it can do so in a non-question begging way. How so? If risks are not symmetrical, the proponent of the riskier position—the position that if one is wrong involves greater harms—bears the initial burden of proof. Suppose the position bearing the greater risk of harm in being wrong is a judgment of permissibility. The proponent of impermissibility enjoys a presumption in favor of her position and does not need to argue for it—initially. Defendants in the court of law, for example, enjoy a presumption in favor of not guilty; the plaintiffs must prove guilt. Call this the burden of proof role. That risk can determine who has the initial burden is compatible with what was said above since the risks in being wrong are, often enough, conceded by disputants.

2. A risk in being wrong about P might make P more sensitive to defeaters. Quite independent of whether the proponent of P bears the burden of proof, one could say that a higher risk in being wrong about P renders P more easily undermined

---

[11] Understanding moral risk in this way is identifying a different problem than what usually falls under the heading of evaluative or moral uncertainty. See MacAskill and Ord (2020), Sepielli (2019), and Harman (2015).

by counter considerations. Consider Braddock's opening case of Brendan Bryant, a drone operator engaged in a missile attack on a high-ranking target.

> Bryant's laser hovered on the corner of the building. "Missile off the rail." Nothing moved inside the compound but the eerily glowing cows and goats. Bryant zoned out at the pils. Then, about six seconds before impact, he saw a hurried movement in the compound. "This figure runs around the corner, the outside, toward the front of the building. And it looked like a little kid to me. Like a little human person"[12]

We are told that in the actual scenario, the operator did have time to veer the missile away from the putative target. Let's suppose that Bryant believes that the drone attack would be rendered impermissible if a child were present in the blast area. Furthermore, suppose he believes that the image supplies evidence, however weak, that a child is present in the blast area. How weak? In other contexts, the presence of such an image would not justify *knowing* that a child is present. What is interesting about the case is that the evidence is enough to justify inaction.[13] The risk of wrongdoing renders the judgment that 'the drone attack is permissible' more sensitive to defeaters, even something as weak as a simple smudge on a low resolution screen. Call this the sensitive to defeater role—or simply the defeater role.

3. A risk in being wrong about P renders one's action on the basis of P immoral (or insufficiently justified). On this understanding, the relevance of risk has less to do with the epistemic standing of my belief that P, and more to do with my moral responsibility to consider the potential harms in being wrong about P. But this understanding can be split into two separate positions. On the one hand, cases where we have the intuition that it would be irresponsible to act on P are used to justify the claim that risks in being wrong about P can render one's action based on P as wrong (Reed, 2012) whether the agent knows P or not. On the other hand, there are cases where we have the intuition that it would be wrong/unjustified to act on P, and this justifies thinking that the agent does not know P to begin with (Fantl & McGrath, 2009). The latter position holds that knowledge is subject to pragmatic encroachment, the former holds that knowledge is immune to pragmatic considerations, but our moral responsibility is not. For both positions—discussed in more detail in Chap. 8—a risk in being wrong about P can render acting on P morally unjustified. In some cases, we might know that P but not enough to justify *acting* on it—acting on it may be deemed careless or negligent. This role aligns closest with Rudner's reflections above. "[O]ur decision regarding the evidence and respecting how strong is '"strong enough"', is going to be a function of the importance, in the typically ethical sense, of making a mistake…" (Rudner, 1953, 2). Risk is a *function* of the gravity of wrongdoing if one were wrong about P in relationship to the justification for P. Call this the bar of justification role, or simply the justification role.

---

[12] See Braddock, 2024, 1; quoted from Power (2013).

[13] The reader is welcome to add the condition that there is no other outweighing good that is secured by continuing the attack.

4. The previous three roles have moral risk affecting in some way the justification for acting. There are a family of risk principles according to which the magnitude of harms that could result from a particular practice are themselves a justification for inaction. There are two ways to specify what risk considerations might be doing in such circumstances. (1) One should avoid performing an action A if and only if, if one's belief B that A is permissible is wrong and one acts on B, a serious wrong would occur; and one's belief B is held with less than absolute certainty. An objective version of this principle would simply change absolute certainty to something like less than an epistemic probability of 1.0 given the entirety of relevant reasons. (2) A second notion is slightly weaker. One should avoid performing an action A if and only if, if one's belief B that A is permissible is wrong and one acts on B, a wrong would occur that is *worse* than acting as if A were impermissible. And belief B is held with less than absolute certainty. (The objective version would be revised as above). Though this is a mouthful the second principle simply involves a comparison. An agent compares the relative outcomes of the following scenarios: (2a) being wrong (unbeknownst to the agent) that A is permissible but acting as if it were permissible; and (2b) being wrong that A is impermissible (unbeknownst to the agent) and acting as if it were impermissible. If the outcomes are morally asymmetrical (i.e., they are worse for acting as if A were permissible), then one should refrain from doing A. For both principles, risk focuses attention on the magnitude of relative harms and justifies inaction on such a basis.

In complicated dialectical exchanges where risk is relevant, there is no reason to think that only one role is functioning. It is not my task, then, to specify which role. But I am sympathetic that 1, 2, and/or 3, is functioning. I have serious doubts about 4 because there are numerous judgments we make and act on them such that if we were wrong about them moral catastrophes would occur daily. Boonin (2003, 314 ff.) supposes that there is a nonzero probability that blades of grass have a right to life. If true, that would make mowing my lawn tantamount to mass murder. But such a moral catastrophe hardly affects the justification of my act of mowing even supposing that letting my lawn grow has few downsides beyond aesthetics.

In general, the notion of moral risk in this book recommends the following: if the moral cost in being wrong in acting on P is greater than the cost of being wrong in acting as if ~P, then the justification for P must be greater than the justification for ~P to justify acting on P. Risk considerations can affect this justification in the ways noted above. By focusing attention on the *justification* for an action, it avoids the counter-intuitive consequences of Boonin's blade of grass example—(and by focusing on the justification for acting and not on belief, it avoids all of Boonin's objections. See Chap. 8 for details).[14]

---

[14] Arguments from risk are not new, and I do not pretend to be comprehensive. A comprehensive account would have to survey the literature on pragmatic encroachment, the precautionary principle, decision making under uncertainty, probabilism vs. probabiliorism, and so on. Specific to the topics canvassed in this project, the menu is more manageable. In no order of importance, see

As with justification-defects, the reasons why an argument suffers from moral risk defects requires comparing and contrasting counter arguments and counter-evidence. I do not hold, again for reasons outlined in Chap. 8, that a risk argument for inaction (or any action) can be determined apart from a detailed analysis of dialectical exchanges.

Such an analysis may look like one is limning de facto defects. The difference is that for a moral risk argument, the goal is to argue that P should not be acted upon because the justifications for P and ~ P are plausibly on par, or that the justifications favor ~P. Of course, arguing that an argument suffers from a moral risk *defect,* requires an evaluative judgment. That judgment might be subject to peer disagreement, as with the claims that the pro-choice position for abortion bears a greater moral risk in being wrong because it involves killing or, conversely, the pro-life position bears a greater moral risk in being wrong because it involves frustrating a liberty interest of the mother. As is discussed in the final chapter, however, there should be agreement on who bears the moral risk. Proponents of abortion rights claim that the risks are wholly on the side of the pro-life view because of a woman's liberty interests. I argue in Chap. 8 that pro-choice proponents cannot understand the axiological landscape in this way given their own commitments. Briefly, the scope of one's liberty interest is circumscribed by concerns about commutative justice. No one thinks that if a mother no longer wants to care for her born child should be allowed to neglect the child. The extent of one's liberty interest is circumscribed by one's just treatment of others. If the other in question is in fact a person (whether born or unborn) the exercise of liberty holds second-place. So, even on the pro-choice view the risk of intentionally killing a person and thus violating one's commitment to commutative justice, is a more important value than the exercise of one's liberty. The gravity of harm or wrongdoing is asymmetrical. Is the probability? To answer that requires inspection of the relevant arguments.

2.4 Conclusion

One should keep in mind that what is said about P applies to any proposition including propositions constituting one's justification (since one's justification include propositions as well). On the surface this might look like a jejune observation but it is important for assessing moral risks. Which moral risks there are is partly described in terms of the moral costs of acting on P if P were false. Since any proposition can

Moller (2011), Grizes (1970), Kreeft (1990), Thomson (1995), Lockhart (2000), Maitzen (2003), Pruss (2011), Beckwith (2007), Boonin (2003), Braddock (2024), Friberg-Fernros (2017). Even though we have a smaller menu here, it is not my intent to hunt for similarities or differences between previous iterations of risk arguments and my own. I'm not interested in being unique, though I do say quite a few things that add to this literature (e.g., the four roles above and the application of moral risk to euthanasia). My main interest is getting the argument correct which will intersect with previous insights.

bear a moral cost in being wrong in acting on it and one's justification is constituted by propositions, two consequences follow. One's justification for what the moral costs there are also bears the same moral risk in being wrong and the justification for attributing moral risk to this or that position can itself bear a moral risk in being wrong.

Concretely stated, a microcosm of the first consequence is as follows. Suppose abortion is defended via the no person strategy (Chap. 4) according to which abortion does not kill one of us. Suppose that 'abortion is permissible' suffers a greater moral risk in being wrong than 'abortion is not permissible'. As is defended in Chap. 4, the no-person argument requires thinking that you and I are essentially psychological entities. And the argument for that claim requires a particular interpretation of brain transplant thought experiments. That interpretation, however, is not the only one that can accommodate our intuitions on transplant cases. I argue in that chapter that *privileging* the interpretation that leads to the psychological view must already assume that you and I are psychological entities (cf. Longino's background assumptions). Such an argument must assume *as basic* the distinction between selves and bodies, or that you and your human animal have different persistence conditions. Since the proposition 'abortion is permissible' bears a risk in being wrong, so too does its justification which, in this example, is a specific understanding of 'self' and 'body' paired with a claim that they are substantially distinct. This justification inherits a risk in being wrong, and so on "down" the dialectic.

Regarding the second consequence, ascribing moral risk to this or that position can be a tricky matter in that it risks de jure defects itself. As with the *Tarasoff* case discussed above, however, if the same values can be held constant, arguments for which position bears the moral risk can itself avoid de jure defects or they are simply not in dispute.

That is just one iteration of how the arguments in this book proceed. I limn an iteration of it here not to spoil the plot but to reveal what is a simple pattern going forward. The procedure is to limn a justification, plume its inferential pattern and background assumptions, and in Chap. 8, collate the results in the form of a risk argument. This pattern is important to canvas here to see it amid rather complicated dialectical analyses of various arguments.

Chapter 3
Impairment Arguments, Interests, and Circularity

3.1 Introduction

Arguments for the permissibility of abortion are divided into two broad strategies: (I) the no-person strategy and (II) the person-but-lacks-feature x strategy.[1] The former category of arguments argue that unborn human beings are not persons and non-persons may be killed. Different arguments are presented for what counts as a human person but for all of them, unborn human beings—especially at the embryonic or zygotic stage—are not persons (Warren, 1973, 56; Warren, 1992; McMahan, 2002; Tooley, 1972, 45 ff.). The latter category (II) of arguments typically grant that unborn human beings are persons in an ontological sense (Degrazia, 2005; Boonin, 2003; Thomson, 1971)—they may grant the truth of the claim "my mother first felt *me* kicking when I was 18 weeks old." But proponents of this strategy argue that unborn human beings lack some morally important feature the lack of which makes it permissible to kill us, even if we are already born (Singer, 1979) or just about to be born (Degrazia, 2005, 290). Two types of features have been proposed: interests (DeGrazia, 2005; McMahan, 2002) and bodily rights (Thomson, 1971; Boonin, 2003). This chapter focuses on arguments from interests, namely, the time-relative interest account (TRIA) of harm or wrongdoing. I reserve a brief discussion of the bodily-rights strategy for the penultimate section and the argument from consciousness for the final section. Chapter 4 considers the no-person strategy.

---

[1] The term 'person' on this second strategy is not understood in any dualistic sense—it is not understood as differentiating between human organism and person. The term 'person' on this second strategy refers to you and me at every point in our existence which includes before birth. Thomson, for example, says, "I am inclined to think also that we shall probably have to agree that the fetus has already become a human person well before birth" (Thomson, 1971, 47). And further on she says, "I propose, then, that we grant that the fetus is a person. From the moment of conception" (p. 48). Famously, the Violinist analogy is meant to show that even a person does not have the right to the use of someone else's body, viz., some persons lack a morally relevant feature that renders abortion permissible.

S. Napier, *Justified Killing*, https://doi.org/10.1007/978-3-032-14946-6_3

Before considering in detail the TRIA argument for the permissibility of abortion, it is necessary to get some idea of what an abortion is in this discussion. In this work, abortion is the intentional killing of a developing human being. The target, aim, or end of the act principally performed by the OB/GYN is to end the life of the developing human being. This intent need not be present for most cases in which a mother's life is threatened by the pregnancy, e.g., ectopic pregnancies. But paradigm cases of a D&C abortion,[2] for example, are cases in which an intent to end the life of the human being is present.

Colgrove (2025), however, presents an interesting case in which a mother rather unfamiliar with embryological development discovers that she is pregnant and procures an abortion pill to remain in her career. She ingests the pill at 7 weeks' gestation and the baby dies. This should count as an abortion, but because the agent does not know enough embryological facts, she cannot *intend* the human being's death. So, intention should not figure in a definition of abortion. Colgrove's solution is to stipulate that an abortion requires that one secure the death of a developing human being (abbreviated "*e*") "without regard for e's survival" (Colgrove, 2025, 161).

Colgrove admits that the clause "without regard for" is conceptually very close to intention but the former tolerates a wider array of killings. He wants a definition of abortion that includes cases involving an intent to kill but also "some acts of feticide that are accompanied by *indifference* to *e*'s survival. This means that abortion may occur even if one is not consciously intending (or desiring) the death of *e*" (Colgrove, 2025, 163).

Colgrove's analysis is percipient, granular, and intuitive. For the following reasons, however, intention should figure in the definition of abortion. First, to not care that a human being survives one must understand that one is present. Satisfying the clause "without regard for" requires the same knowledge that having an intention requires. So, Colgrove's definition requires knowledge of developmental facts just as much as a definition in terms of intention does. Second, a doctor managing an ectopic pregnancy might lack regard for the embryo's survival—because she simply *cannot* make it such that the embryo survives. She might be "indifferent to e's survival." But salpingectomies (procedures which can correct an ectopic pregnancy) are not considered abortions. Third, Colgrove's case of the mother unthinkingly procuring an abortion pill is still a case of abortion even on a version that requires intent. She performs an act, and therefore, has an end in mind. What is that end? She might think falsely that it is a mere clump-of-cells, but she clearly does not want whatever it is to live anymore. She intends the death of a *human being* and this is not a surreptitious insertion of "human being." When defining action-types, some abstraction is necessary. A perpetrator of domestic violence, after all, may understand (and thereby intend) only to keep his partner 'in line' but he performs acts of battery just the same—he intends bodily harm. Likewise, someone might understand an abortion only to involve the death of a clump of cells, but law and moral

---

[2] For a description of abortion procedures, see Dr. Anthony Levatino, "Abortion Procedures: 1st, 2nd, 3rd Trimesters." https://www.youtube.com/watch?v=CFZDhM5Gwhk&t=213s (accessed 19 August 2025).

analysis understands intention according to actual fact. Finally, paradigm instances of abortions are performed by an OB/GYN doctor, and there is no lack of embryological fact that could preclude him from intending the death of a developing human being. Borderline cases are those in which the mother suffers from, for example, pulmonary hypertension, or severe diabetes—both of which are conditions that can be severely exacerbated by a pregnancy. We can safely set these cases aside as the typical arguments for (pk) are never so narrowly circumscribed.

We may now consider the TRIA justification in detail. The basic idea is that an abortion does not violate any of the fetus's time-relative interests. The time-relative interest account of harm and wrongdoing tells us that a necessary condition for harming someone is that his or her time-relative interests are frustrated. Regarding the justification for abortion, this account falls prey to impairment arguments. Impairment arguments entertain cases of prenatal injury, such as the mother using illicit drugs that disable the child. The intuition is that the child who is born with such disabilities is harmed by the mother's drug use. But it is unclear what time-relative interest is violated in cases of prenatal harm—assume also *preconscious* harm. Typical responses to impairment arguments point out that the abortion case is different because the child does not exist to experience such harms; but in prenatal injury + survival cases, the child does live to experience those harms. Thus, the TRIA justification for abortion is not impugned by impairment counterexamples. This chapter argues that this response to impairment arguments is viciously circular. The response must say that so long as you kill the child, no harm is done. But this assumes that killing itself is morally inconsequential and is not itself a case of harm. The response to impairment arguments, then, assumes the permissibility of abortion.

To see this pattern in higher resolution, suppose I believe that the Earth is flat. You point out that ships that sail away slowly pass out of sight as if they are going down. If correct, that visual presentation would be inconsistent with the Earth being flat. I reply that the electro-magnetic field of the Earth distorts light waves to make objects emitting those waves look like they are going down. Flummoxed, you ask why I believe that such electro-magnetic waves exist, let alone that they exert the visual effects I attribute to them. I respond that a flat Earth produces such electro-magnetic waves (and I offer no independent reason for this claim). Frustrated, you point out that I am providing a circular justification. I am using the conclusion of an argument to deflect a claim against the premises of my very argument.

This chapter argues that the TRIA justification for abortion rights must recapitulate a similar circular pattern. What is so wrong about such circular reasoning is explored in Sect. 3.4. Specifically, the present chapter argues that an argument for abortion rights based on a time-relative interest account of harm is circular—an iteration of an inference-defect. The important development to note is that the present argument does not rely on views regarding personhood or bodily rights.

The second section describes the key concepts of the TRIA of harm or wrongdoing, and how it justifies the permissibility of abortion. The third section explains and defends the impairment argument and demonstrates that TRIA must respond to it in a circular fashion. The fourth section explains why the circularity is a serious problem for the TRIA justification of abortion. The basic pattern of my argument is as follows. Start the dialectic with a view on what counts as harming or wronging someone. This view purports to justify that killing pre-conscious human beings is permissible (I abbreviate this position as "permissible killing" or (pk)). There are cases, however, where it looks like pre-conscious human beings can be subjects of harm or wronging. To deflect such putative counter-examples, the defender of (pk) must provide a rejoinder. The bulk of my argument is to argue that this rejoinder is justified only if one assumes that (pk) is true. The fifth section entertains an important objection to the argument in this chapter. Briefly, the proponent of the TRIA account can say that the TRIA is the best option out there, and no axiological view is free of philosophical cost. So, even though there is a circular pattern of justification for (pk), it is still the best view available. I show that the "philosophical cost" referred to here can only be that other axiological views are compatible with abortion being impermissible. The circularity recapitulates at a higher level of justification.

3.2 The Time-Relative Interest Account and the Justification for Abortion Rights

The central insight of the TRIA is that it serves as an explanation for why certain states of affairs (SOA) can be the object of rational, egoistic concern. What matters to subject S is whatever S can take an interest in, and this is a function of two factors. The first is the importance to S in realizing (or avoiding) an SOA, and the second is the degree of psychological unity between S now and when S realizes (or avoids) the SOA in the future. Having a time-relative interest requires meeting both conditions.

> The extent to which one ought now to be egoistically concerned about that event is a function of two factors: first, the value, positive or negative, that the event would have at the time when it would occur, and second, the extent to which the prudential unity relations would hold between oneself now and oneself at the later time... (McMahan, 2002, 79; see also DeGrazia, 2005, 191).

One must add a third necessary condition, namely the capacity for having time-relative interests at all, which includes the capacity for consciousness, sentience, and memorial capacities.

Psychological unity is a function of having the same brain, understanding 'the same' as physical, functional, and organizational continuity—i.e., the preservation of those brain areas associated with consciousness (McMahan, 2002, 66–69). It is not necessary to remember everything I desired, say, 10 years ago to be psychologically unified with myself 10 years ago. All that is required, on McMahan's view, is an overlapping psychological unity which is anchored in having the same functional brain.

TRIA proponents also think that what explains rational, egoistic concern can also explain what counts as a harm. Hitting a tennis ball with my racket is not morally charged, hitting my cat with my racket is; the former does not have any interests, the latter does. More technically, if there is an SOA for which there are weak or non-existent psychological unity relations between me now and the SOA in the future, I cannot be harmed by being deprived of that SOA. If a toddler cannot foresee, intend, or understand having chocolate cake for breakfast, not serving him the cake cannot matter to him. From his own perspective, he misses nothing by having Cheerios instead. Thus, if an SOA does not *matter* to me, I am not harmed by not experiencing it. From these admittedly plausible reflections, the TRIA makes the following inference: if it is not me who will survive into the future or the psychological unity relations are weak—as is the case between an early developing human being and an adult—survival itself cannot be an object of rational, egoistic concern—survival cannot matter to me. Missing out on living would be like the toddler missing out on cake. That might sounds crass, but what makes the TRIA plausible is that it focuses on what matters to the subject; read, *egoistic* concern. And from that perspective, if I cannot conceive of surviving or living, it cannot matter to me either.

The idea of "mattering" is used to describe the "concern" part of a rational, egoistic concern. There is an informative objection-reply that reveals a deeper understanding of TRIA relevant to the impairment arguments discussed below. If we interpret the question "what matters to subject S?" as asking what, from the first-person perspective, is understood by the agent herself as being an important SOA to realize, having a time-relative interest seems essential. If we interpret the question "what matters to S?" as asking what would redound to S's benefit understood more objectively, having a time-relative interest is not essential. If we consider the latter, one might harm someone who is not conscious of the harm or cannot understand it as a harm. The reply is that it is hard to understand how S is harmed (or benefited) by SOA's for which S cannot or will not ever experience. An unexperienced harm, as far as S is concerned, is unintelligible on the TRIA. Though I disagree with this claim, more on that later, it has a ring of plausibility to it: an *unexperienced* harm is no harm at all.

The TRIA of rational egoistic concern can function as an argument for the permissibility of abortion and such an argument would not have to make assumptions about personhood. Consistent with category (II) arguments, suppose that you and I are ontologically identical to the developing human being in utero; the TRIA claims that because the psychological unity relations are so weak or non-existent, there is no harm to the fetus if she or he does not survive. She is like the child who is deprived of chocolate cake for breakfast—she misses nothing.

DeGrazia's argument for the permissibility of abortion, for example, focuses on the moral relevance of interests according to which the pre-conscious child[3] does

---

[3] I use the term 'pre-conscious child' because DeGrazia grants our numerical identity from the time after twinning is possible and onward. McMahan too thinks that late-term abortions are permissible even after the onset of consciousness—a criterion for personal identity on his embodied mind

not have the requisite interests to ground a claim that she or he can be harmed. DeGrazia notes that our intuitions on harm are parasitic on "the way in which one is psychologically "invested" in, or connected with, one's future" (DeGrazia, 2005, 286). Consistent with what is noted above, the notion of interest here is that one must be able to *take* an interest in continued living, which implies having an actual interest.[4] But this "taking" is a degreed concept. DeGrazia states that "even if the presentient fetus has an interest in remaining alive, it would be too weak to ground a right to life" (2005, 288). It is too weak either because the fetus cannot form an interest given that her or his cognitive and conative abilities are too incipient, or one's memory capacities are too incipient to ground any *connectedness* between psychological states.

So, there is a triad of concepts; wrongdoing, harm, and interests that are related roughly as follows. Wrongdoing requires harming and harming requires violating someone's time-relative interests. Having an interest requires being conscious—or having enough mental capacity to take an interest in something. Since preconscious human beings cannot take an interest in continued living, ending their lives cannot violate or frustrate any of their time-relative interests. Therefore, pre-conscious human beings cannot be harmed. If they cannot be harmed, killing them is permissible. Therefore, killing them is permissible.

3.3 Impairment Arguments[5]

Impairment arguments are arguments against the time-relative interest account of harm, according to which there are cases where a subject is clearly harmed by another's actions, but, intuitively, there is no time-relative interest that is violated or frustrated. Specifically, impairment arguments focus on putative harms on preconscious human beings. The idea is that if such harms are possible then the fact that one cannot violate a time-relative interest is still not sufficient to justify permissibly killing preconscious human beings. That is, if impairment arguments are successful, they justify the following claim:

> (Impairment): If it is possible to harm preconscious human beings, then it is not necessarily the case that if there is no time-relative interest that is violated by such an action, it is permissible to kill that preconscious human being.

---

view. McMahan's argument for the permissibility of late term abortion is that the prudential unity relations are too weak. Both think that the absence of a rational egoistic concern for the future is sufficient to say that the agent cannot be harmed. Understood in this way, the TRIA justification for abortion rights is meant to prescind from metaphysical questions of personhood.

[4] I do not argue that TRIA is false but there are plausible arguments to this effect, see Tollefsen (2011), Holtug (2011), and Nichols (2012).

[5] Sections 3.3 and 3.4 are largely from Napier (2024).

The idea behind impairment arguments is that if harm is possible *sans* violating a time-relative interest, then killing that preconscious human being *could* count as a harm.

The following are some putative counterexamples to motivate impairment arguments. Consider prenatal impairment according to which an agent harms the pre-conscious child—for example, the mother abuses drugs voluntarily while pregnant with the child (DeGrazia, 2005, 289; McMahan, 2002, 280; see also Montague, 1989). The intuition is that the mother does something impermissible in harming the pre-conscious child. But if that child does not have or has very weak prudential concerns for the future, how can we explain the intuition that the mother's action is wrong just on TRIA?

Ingmar Persson (1999) describes a case of neuron removal to illustrate the possibility of preconscious harm as well.

> Consider a series of removals of neurons from the brains of preconscious fetuses… Imagine that this piece of surgery could be performed without killing the fetuses, but that it would cause them to be mentally retarded to some degree. If one removed just a few neurons, this would harm the fetus…If it would restrict its mental life and thus make it less capable of leading a worthwhile life. If one removed still more neurons, this would harm the fetus more were it to make its future life even less worthwhile… Suppose, finally, that so much of the fetus's cerebrum is removed that it never acquires any consciousness at all…Then, according to [a view that harm requires consciousness], this would not harm the fetus (1999, 301).

If having future experiences is a necessary condition for being harmed, then it looks as if the preconscious child is harmed by neuronal removal *except when* that removal undercuts the possibility of developing consciousness at all. However, Persson observes correctly that this ultimate manipulation inflicts a "*greater* harm on the fetus than the penultimate manipulation did…" (1999, 302).[6]

Lastly, consider a case of preterm injury plus restoration. Suppose a preterm mother suffers domestic abuse and as a result, the preconscious child is smitten in the womb such that he suffers severe brain damage. Suppose that there exists a fetal brain surgery to correct the damage and in this case is completely restorative. In this case, the child is harmed *by the battery*, for if not, there is no sense in which we can say that the surgery was *restorative* or even that the surgery is *needed*. Those descriptors make little sense without supposing that an injury has occurred. This case suggests that harms can occur to an agent without that agent experiencing the effects of those harms.

---

[6] See Hendricks (2019a, b) and Blackshaw (2019) for important discussions on harms that come in progressive stages. At issue for Blackshaw is that the ultimate manipulation, the one that kills, does not inflict a *greater* harm. Even if Blackshaw is right, impairment arguments may still succeed if they can show that one can harm a preconscious human being however severely.

3.3.1 Response to Impairment Arguments

DeGrazia correctly notes that such actions are wrong because the pre-conscious child has a "present time-relative interest in being healthy but also many future time-relative interests in being healthy…" (2005, 289). DeGrazia also correctly recognizes the problem saying, "Since the presentient fetus [in the mother-drug case] is psychologically cut off from the child he will become, his time-relative interest in later being healthy is extremely weak, just as his time-relative interest in staying alive is extremely weak. How, then, to explain the judgment that the woman's behavior is seriously objectionable…?" (2005, 289).

It would seem to follow that the pre-conscious child who is killed by an abortion would also have a present time-relative interest in being healthy. If the drug abused child is harmed and harming requires having interests, the drug abused child must have interests that are violated. If the drug abused child can have such an interest, so may the child who is killed by an abortion since we may easily assume that they are developmentally similar vis-à-vis having interests. And so, it would follow that killing the child involves a significant harm as well. So, if the preconscious child is harmed by prenatal impairment, the preconscious child who is killed is also harmed. Both are wronged.

All the authors who consider impairment arguments agree that in cases where the child *survives* prenatal harm, the child is wronged. There is only one option open to the TRIA justification for permissible killing (pk). The impaired child is harmed but the aborted child is not because of further specifications on what count as morally relevant interests. If we stick to what proponents of the TRIA say, those specifications take the following two forms.

(1) *Experience condition.* In non-abortion harm cases, the child survives into adulthood and experiences the effects of impairment (e.g., the mother's drug habit or neuronal removal). The drug case is immoral because the child *will* exist into the future and *experience* the harms that were originally caused by the mother's drug habit. It is only *at the point of experiencing* the harms that one can say the person is harmed. DeGrazia intimates this option when he comments on the drug abused child: "*Because the woman decided against abortion*, her fetus has not only a present time-relative interest[7] but also many future time-relative interests in being healthy…" (2005, 289). And for the aborted child, he states

---

[7] In saying this, DeGrazia seems to grant that the preconscious child at the time she is still preconscious has a present time-relative interest. Taking an interest seems to require having desires or at least some psychological capacity. So, it is implausible to suggest that a preconscious human being can also take an interest. Granted, he does say that the present interest is weak. But weakness on the TRIA is a function of psychological *connectedness* to one's future (or past) psychological states which requires functional memory (McMahan, 2002, 39 and 43). One could say that what matters here is having an objective interest, for example, a blade of grass wants to grow but is not conscious and does not technically take an interest in growing. But the moral relevance of setting back an objective interest is questionable. It is not immoral for me to mow my lawn even though I'm setting back the objective interests of my grass wanting to grow. Resorting to objective inter-

that "...we count only the fetus's present time-relative interest to live, because an aborted fetus *will never have time-relative interests* while psychologically deeply unified..." (2005, 289, emphasis added). When explaining the TRIA, DeGrazia says, "[t]he essential idea is that prudential evaluation of a possible future...should take into account both the value of that future to one *as one experiences it* and how psychologically invested one now is in that future" (2005, 191, emphasis mine). Likewise, McMahan observes that "abortion affects its victim only when the victim is a fetus with weak time-relative interests, prenatal harm affects its victim *later...*" (2002, 282, emphasis mine). McMahan thinks that the moral relevance of this distinction is simply that the fetus in the abortion scenario will not have any future time-relative interests, because she or he is killed[8]; but if that abortion is not performed, the fetus will have time-relative interests. And if the aborted child cannot have any future time-relative interests, she or he cannot be harmed. Blackshaw (2019, 724), following McMahan (2006) adds that "killing the fetus entails that there is *no* future individual with interests to be damaged. Consequently, the *reason* why [inducing fetal alcohol syndrome] to a fetus is immoral does not apply to the act of killing the fetus." So, for S to be harmed, S must experience the harm. Since the aborted child does not experience anything, neither actual nor future interests are frustrated and consequently, no harm either.[9]

(2) *Content specification.* The second option is to specify the content of the time-relative interest. DeGrazia suggests this option when he supposes that in the abortion scenario "we count only...the present time-relative interest *to live...*" (2005, 289, emphasis added). The idea seems to be this: both the aborted and drug-abused child cannot form an interest in continued living, since that requires having a self-concept, and projecting one's self as existing into the future. With no interest, no harm. But the drug-abused child has an interest in being *healthy*, albeit a weak one.

Content specification holds that the impaired child is harmed *at the time* of the drug abuse or neuron removal but that the aborted child does not form an interest *to live* and so *that* interest cannot be violated. And the Experience condition holds that the

---

ests, then, would not explain why we think that the drug abused child is wronged *at the time of the injury*. Given (1), wrongdoing seems to occur later.

[8] Gillham (2021, e43) echoes similar sentiments saying that, "[s]ince fetuses that will not be born do not have futures full of experiences, activities, projects and enjoyments to render their lives valuable, fetuses have no [future like ours]. Therefore,... aborting them does not deprive them of an [future like ours],"

[9] The reason for describing this as the *experience* condition and not, say, as the future *interest* condition is because some forms of harming can render the formation of interests in the future rather difficult or impossible, as with progressive neuron removal. But we still think that if such a being lives to experience those harms, that is enough to ground claims of moral wrongdoing. If I were to focus too narrowly on interests being violated specifically, it would be too easy to generate counter-examples to TRIA since it is not hard to imagine that the child suffering from fetal alcohol syndrome fails to *ever* form an interest in continued living or in being healthy.

impaired child is harmed *only when* the effects of the drug abuse or neuron removal can be experienced. The Experience condition tells us that neither the aborted nor the pre-natal impaired child is harmed at the time that the putative harms are caused. If the impaired child were to die of Sudden Infant Death Syndrome prior to experiencing the harms of drug abuse, there still is no wrongdoing given the Experience condition.

Of course, the normal default attribution of wrongdoing is that it occurs when the injury occurs. In the battery case, for instance, the child was wronged at the time of battery. Consider a case of unethical gene therapy research according to which the effects will not be realized for another few years. Intuitively, researchers performed a wrong act *at the time* they did, say, a risky gene therapy infusion, but the experience of the harms is displaced in time. The proponent of the Experience condition will, however, put emphasis on the fact that *as far as S is concerned*, unexperienced harms do not matter to her. And if they cannot matter to S, S cannot be wronged.

3.3.2 Why the Response is Circular

I can now argue that the TRIA justification for (pk) is circular. Impairment arguments function as an undercutting defeater to TRIA's justification for (pk). Prior to considering impairment arguments, the TRIA justification has it that if S lacks the ability to have a time-relative interest in continued living, S cannot be harmed. Impairment arguments suggest that harms can occur even in the absence of such time-relative interests.

Consider the Experience condition first. This condition functions as a defeater *deflector* to the impairment argument thereby preserving TRIA's justification for (pk). The upshot of that condition is that we have two classes of cases: prenatal harm plus survival, and prenatal killing (i.e., abortion). The only apparent difference between the two classes of cases is that the child is killed in the abortion case, but not in the survival case—we can imagine that they are developmentally the same. Grounding a moral asymmetry between these cases must say that *killing itself* is the difference maker such that the child in the abortion case is not harmed, but in survival cases, the child is harmed. So long as one kills preconscious human beings, it does not matter how many neurons are removed, or how many parts of the brain are damaged from battery or illicit drug use.[10]

This claim *assumes* that killing is not an instance of harming. The abortion doctor does not do anything wrong in the abortion case because the child is neither

---

[10] One reviewer noted that an implication of the TRIA at this point is that in cases in which a child survives an abortion but with serious injuries, one is *obligated* to kill the child to prevent the child from experiencing the effects of those injuries. The TRIA paired with the Experience Condition delivers this rather counter-intuitive consequence.

conscious (of the harms) nor will she ever be.[11] If killing itself is the difference maker, which entails that the child is not wronged if she or he is killed, then deflecting impairment arguments must assume that killing is permissible. So, to deflect impairment arguments, the TRIA justification for (pk) must already assume that killing preconscious human beings is morally permissible. The so-long-as-one-kills response amounts to saying that those counterexamples pose no threat in the abortion case because *killing preconscious human beings is morally permissible*. The background assumption that adjudicates whether various types of prenatal harm cases count against the TRIA is the very claim that needs argument, namely, (pk). Therefore (pk) is presupposed to deflect a defeater to a premise in the argument for (pk). Such an argument is circular in that the claim that needs defending, i.e., (pk), is being presumed.

Consider now the Content Condition. If what makes it immoral to harm the child in the first case is that the child has a present time-relative interest in being *healthy*, then the same interest is sufficient to say that *killing* the child counts as a harm in so far as not existing is incongruent with being healthy. The issue here is not that existing is merely a necessary condition for being healthy. My claim is that if S were to have an interest in being healthy, that interest entails existing regardless of whether S forms the latter interest explicitly.[12] Suppose a vapid teenager forms an interest in being healthy but never explicitly makes the inference that he must exist as well. He still enjoys a right not to be killed. If it is an *ability* in taking an interest in continued living that matters, the conceptual content of 'being healthy' is no more complicated to form than forming an interest in living.

Furthermore, why "count only" the interest that happens to preserve the belief that abortion is permissible? If the 'count only' clause is there *so that* permissible killing is preserved from counter-example, then we have an instance of circular justification again. If it is based on independent reasons, what could they be? DeGrazia does not say.

To summarize, the only reason for intercalating deflectors to counter-examples of preconscious wronging is to make consistent one's commitment to: (a) an interest account of harm/wrongdoing, (b) one's belief that *some* cases of prenatal impairment are wrong, and (c) one's belief that abortion is permissible. When faced with a putative example of preconscious harm, one can either accept that harms and interests are not coextensive such that harms can occur before interests form. On this option, one loses a justification for abortion strictly on TRIA. Or one can argue that contrary to appearances, they are coextensive. Arguing that they are coextensive

---

[11] I think that the first disjunct is false, but I am assuming it here for the sake of argument. See Howsepian (2011), Derbyshire and Brockman (2020), and The endowment for human development, "Response to touch," (available at https://www.ehd.org/movies-index.php (accessed 21 June 2024)). The latter provides evidence of sentience as early as the sixth week of human development.

[12] Watt (2022) argues more directly for the conclusion that the pre-conscious human being has an interest in living because having an interest in objective benefits does not require psychological connections. It is not counter-intuitive to say that a tree, for example, has an interest in sunlight, or in not being cut down.

contrary to appearances, requires (1) or (2) as further specifications. The Experience condition presupposes (pk), and therefore the TRIA justification for (pk) is circular. Content specification is independently implausible.

Before transitioning to the next section, it is important to note that my argument for vicious circularity does not require surveying all the intuitions that serve to motivate the TRIA. My argument is simply the claim that in deflecting a defeater, namely, endorsing the Experience condition, the TRIA justification for (pk) must presuppose (pk) itself. This is a structural defect; it is an epistemically circular argument. In any case, as is discussed below, surveying our intuitions on other cases yields underwhelming results for TRIA.

3.4 Why is this Circularity a Problem for the TRIA's Justification for (pk)?

Even if my argument is correct that the TRIA argument for (pk) is circular one might complain asking cynically, "so what if they are circular?" Arguments for the reliability of one's cognitive faculties are circular (Alston, 1993), but we do not infer global skepticism over that fact. A related objection is that the argument from TRIA to (pk) is, at least, a self-contained coherent system. Though it cannot prove alternatives false, it is a coherent system of mutually supporting beliefs. Coherence is not a reason for skepticism.

The first reply that comes to mind is that such a response is hypocritical. For instance, Bigelow and Pargetter say in their critique of pro-life arguments the following, "[b]ut now the issue is to justify, in a manner which *does not beg-the-question*, the claim that the conceptus does have potential in this sense to be a person,…" (1988, 178–79, emphasis mine). Boonin also endorses similar epistemic standards in his critique of pro-life arguments. "I've been asking whether there is a reason that can be given for making species membership morally relevant, a reason that could be grounded in some other belief or set of beliefs *that critics and defenders of abortion both are likely to share*" (Boonin, 2003, 27, emphasis mine). Further on, Boonin remarks in response to a different pro-life argument that "while this claim may turn out to be true, it simply is not plainly true. And *in the context of the debate about abortion … it cannot reasonably be simply assumed to be true…*" (Boonin, 2003, 53, emphasis mine). And on the next page he reiterates the epistemic standard he is assuming stating it as follows. "I am concerned in this book to examine those arguments with which a critic of abortion *can attempt to convince those not already committed to the thesis that abortion is morally impermissible*" (Boonin, 2003, 54, emphasis mine). Since pro-life persons are not already committed to the permissibility of killing pre-born human beings, the TRIA justification for abortion turns out to be a bad argument on Boonin's and Bigelow & Pargetter's standards.

Claims of hypocrisy are rhetorically advantageous, but they are not where I wish to put emphasis since they bear an uncomfortable propinquity with *tu quoque* arguments. The more important point of the above quotations is that they reveal an important agreement on norms of discourse. Norms of discourse, particularly in moral philosophy, require giving and responding to reasons. This is especially true in settings of peer disagreement according to which there exists plausible challenges to one's conclusions (Lee, 2010; Beckwith, 2007). In such a setting, proponents of such conclusions are understood to *give* reasons and *to justify* their position. These norms of discourse are endorsed by my interlocutors, per the quotations above. Do circular arguments succeed in meeting these norms?

Those who argue that circularity is benign typically do so on two grounds. The first is if there are independent reasons for the conclusion of a circular argument (Walton, 1985, 265). The second ground is if there is no other way to argue for a claim except in a circular pattern. The first is benign for obvious reasons. The second ground occurs mainly in discussions on proving the reliability of one's cognitive faculties. The circularity is clear: any argument for the reliability of one's cognitive faculty F must use that very faculty as an argument in support of its own reliability—because it must argue based on a track record of generating true beliefs. The conclusion of such an argument, 'F is reliable' must be assumed in defending the premises such as, F produced belief B *and B is true*. I can only defend that *B is true* if one assumes that the faculty producing the belief is reliable. What is more, if the skeptic thinks that arguments for the reliability of one's faculties are circular and circularity is epistemically bad, then all our faculties are called into question including the faculty that generates the skeptical argument itself. The skeptic, then, risks self-defeat. So, settling for circularity when one must do so is benign.

In the present discussion, neither ground is applicable. Any reason for thinking that (pk) is true independent of the TRIA would no longer be an argument based on the TRIA. And denying either (pk) or TRIA does not lead to self-defeat. The reason circularity infects proofs for the reliability of one's faculties is because we cannot but think, perceive, or generate proofs without using those faculties. Believing (pk) or TRIA is not a pre-condition for thinking the way believing in the reliability of one's own faculties is a pre-condition for thinking.

So, features that would make circularity benign do not apply to the circularity infecting the TRIA justification for (pk). Is there a positive reason for why the circularity is vicious? Most commentators on epistemic circularity agree that what makes epistemically circular arguments vicious is that they fail *to justify* their conclusions in actual dialectical encounters. Walton explains as follows:

> [T]here may be one special kind of context where a circular argument can be seen to violate a reasonable procedural requirement of good dialogue. This context occurs where it is acknowledged by both players that each must prove or argue from premises that the other excepts as *more plausible* than the conclusion each prover is supposed to establish. (Walton, 1985, 271).

Walton accepts that circular arguments might be formally valid, but in these special contexts, they do not provide a ""useful inference" if the premises are as doubtful as the conclusion" (Walton 1985, 271–272). Walton later provides two criteria that circular arguments must satisfy, the first of which is that there is "a circular sequence of reasoning, where the conclusion to be established is either identical to one of the premises, or the premise in question depends on the conclusion,..." (Walton, 1991, 11). The second is that the circular reasoning is used in a dialogical context giving reasons to one's interlocutor who is initially agnostic or non-committed to the conclusion in question. The abortion debate is such a context in which interlocutors give and respond to reasons.

Cling (2003) extracts similar motifs in his treatment of circularity, focusing specifically on the dialectical role of argumentation. On his terminology, justification-affording arguments for a claim C give one's target audience justification for believing in C. Within the class of justification-affording arguments there are those that are justification-creating according to which such arguments "create justification for belief in their conclusions" (Cling, 2003, 281) specific to one's target audience. And there are those that are justification-enhancing according to which such arguments "enhance the justification that their target audiences already have for believing their conclusions" (Cling, 2003, 281). Epistemically circular arguments cannot be justification-creating however. "For an argument P therefore C is justification-creating for audience S only if it is such that *if S were not justified* in believing that C, then S *could* come to be justified in believing that C by reasoning through P to C" (Cling, 2003, 287, emphasis mine). Epistemically circular arguments (self-supporting arguments on Cling's terminology), however, cannot create justification. "For in such a case S's justification for believing if P, then C depends upon S's justifiably believing that C" (Cling, 2003, 287). So, S would have to justifiably believe C *already*, in which case circular arguments are not justification-creating. S could not *acquire* justification for C if C is already presupposed. Circular arguments do not give one reasons to acquire new beliefs or to change the ones they already have.

Sgaravatti argues that circular arguments are bad for very similar reasons, namely, they are dialectically inert, they give one's interlocutors no reason to acquire new beliefs or change the ones they have. Sgaravatti explains that through circular arguments

> ...you cannot *acquire* a justified belief in their conclusion, assuming that you cannot acquire a justified belief by inferring from unjustified premises...whenever the conclusion is actually doubted or is believed without justification, we cannot improve on that situation by making use of the [circular] argument (Sgaravatti, 2013, 771).

Alston too voices these same concerns about the force or usefulness of circular arguments.

> The reason why epistemic circularity is important in this context is that arguments that are infected with it would seem to have *no force*. If we have to assume [belief in the reliability of perception (RP)] in order to be entitled to premises for an argument for it, how can the argument provide support for RP? If our assumption of RP is warranted before we give the argument, how does the argument *add* to the [positive epistemic status] of RP? (Alston, 2005 202, emphasis mine).

Finally, Bergmann notes that circular arguments can be benign when the premises of such arguments are justified non-inferentially, viz. they are first principles. He explains how they can be bad as follows. "The sort of belief that can get infected with epistemic circularity is a belief that one's belief source, X, is trustworthy. A context in which epistemic circularity is a bad thing is one in which the *subject begins by doubting or being unsure* of X's trustworthiness" (Bergmann, 2004, 717, emphasis mine).

The common theme here is that circular arguments are bad when a conclusion or premise in those arguments is questioned, doubted, or is directed to an audience that doesn't already believe the premises or conclusion. In such contexts, circular arguments are inert or dialectically useless; they do not epistemically move the interlocutors doxastic commitments. They are not justification-creating in Cling's terminology; or they fail *to justify* their conclusions in Alston's terminology.

We can now see why the circularity which infects the TRIA justification for (pk) is vicious. The commentators above agree on one aspect of circular arguments, namely, they are bad in dialectical contexts in which the proponent of the argument needs to prove or justify a claim to an audience other than the proponent himself. It is precisely this activity of justifying a claim that circular arguments under perform. In the abortion context, what needs justifying is that it is permissible to kill a developing human being who is ontologically identical to you and me (DeGrazia, 2005), or a person as understood in Thomson's argument (1971), and McMahan's argument (for late term abortions (2002)). The proponent of such a position is saddled with the epistemic project of creating justification for that claim. I take it as uncontroversial that the discourse on abortion requires giving reasons and justifying one's beliefs. Therefore, offering justification-creating arguments is the goalpost.

The reason why this is the goalpost is because my interlocutors agree with such norms. Even if they did not, however, there are independent reasons for thinking that providing justification-creating arguments is required. One important reason is that there exists *peer* disagreement about the merits of (pk). In the settings of peer disagreement, it is not enough to rely on justification-enhancing arguments. Standards of premise acceptability typically require meeting challenges from one's opponent. Freeman explains the problem of accepting premises simply because they form a coherent whole. "Of course, one condition making a premise acceptable would be it's being adequately defended by cogent argument" (Freeman, 2005, 19). This is not a satisfactory condition on Freeman's view since,

> [i]f the defending argument is cogent, *it's* premises must be acceptable and adequately connected to justify accepting the conclusion. Barring circularity, we must eventually come to basic premises, the starting points of reasoning. Under what circumstances, if any, are *these* premises acceptable without any further justificatory argument?" (Freeman, 2005, 19).

Acceptable premises, on Freeman's view, are those premises that can be presumed and presumption is defined with reference to what a challenger must concede, not to mere coherence. In the setting of peer disagreement around (pk), (pk) is not an acceptable premise, and some find it positively outrageous. Therefore, it requires further support. That premise cannot function as a reason to rebut the impairment objection.

To summarize, the reasons why the TRIA argument for (pk) is *viciously* circular is that it (i) violates norms of discourse that my interlocutors endorse. (ii) Mere coherence fails to satisfy basic conditions of premise acceptability in the setting of peer disagreement. The defects here are both inference defects (specifically, circularity) and an inquiry defect.

3.5 Objections: TRIA is the Best View; Other Views Are Too Philosophically Costly

One may grant that TRIA's justification for (pk) is circular and circularity is bad, but still reserve the epistemic right to endorse such a justification anyway. As I see it, the only way to argue for this is to argue more generally that the philosophical costs of holding TRIA are acceptable *relative to other alternatives*.

What we find on closer inspection is that when the comparisons are made either one of the following results occur: (i) The TRIA delivers judgments on cases that conflict with our considered intuitions. (ii) The proponents of the TRIA reject other theories because they do not accommodate our considered intuitions (on TRIA's favored cases). (iii) The specific intuitions that the TRIA proponents say other theories cannot accommodate include (pk) itself. I consider these results in turn, but their implications should be clear: Claim (i) entails that the TRIA inherits philosophical costs. Claim (ii), if true, entails that the selection of which view counts as *best* is based on a prejudicial methodology. Claim (iii) entails that the preference for TRIA as a justification for (pk) is obviously circular.

3.5.1 TRIA and an Inclusive Representation of Our Considered Intuitions

As discussed in Chap. 1, the first step in executing reflective equilibrium (RE) is to collate one's initial considered judgments, specifically our intuitions on cases involving putative harm and wrongdoing. The second step is to discern the incidence with which these cases involve a violation of interests, time-relative or otherwise. The third step would be to manage any adjustment decisions.

We can combine step one and two by collating our intuitions on putative cases of harm/wrongdoing in relation to the property or feature of interest. So, there are logically four classes of cases:

Class I cases are those for which S is harmed/wronged, and S's interests are frustrated or setback. An example populating this class might be hitting my cat with my tennis racket.

Class II cases are those for which S is not harmed/wronged, and S's interests are not frustrated or setback (and this *could* be because S cannot yet *take* an interest in something of value). An example populating this class might be hitting a tennis ball with my tennis racket.

Class III cases are those for which S is not harmed/wronged but S's interests are frustrated or setback. An example populating this class might be justified paternalism.

Class IV cases are those for which S is harmed/wronged and S's interests are not frustrated or setback. An example populating this class *might* be cases of prenatal impairment.If there are cases satisfying the criteria of Class IV, the TRIA is disconfirmed. So, I focus attention on whether Class IV is populated first.

Consider the following five cases as putative members of Class IV cases.

(a) *Research ethics abuses.* Suppose I suffer a serious adverse event in a research study for gene therapy the full consequences of which may not be revealed for another few years. (Suppose also that this was a case of negligence by the researchers.) I am harmed and wronged *by that negligent act*, even though I do not experience the harm at the time that the wrongful act was performed.

(b) *Battery plus surgery.* Suppose the mother suffers domestic abuse and as a result the preconscious child is smitten in the womb such that she suffers severely damaged vertebrae. Fetal surgery is done a few weeks later and it is completely restorative. In this case, the child is clearly harmed by the battery, but does not experience any of the effects of that battery.

(c) *Battery plus death.* Suppose that the mother suffers domestic abuse that causes the preconscious child to die. Here too, the child is harmed but will never experience the harms. That the child is killed cannot be the moral difference maker between this case and the previous one without already thinking that killing is morally inconsequential.

(d) *Rape of an unconscious patient.* Suppose that an unconscious patient is raped by one or several of the clinical staff members. Here is a case where the patient is wronged but does not experience the harms. This intuition survives even if we suppose that the unconscious patient never regains consciousness.

(e) *Rape and killing of an unconscious patient.* Suppose that, as in the previous case an unconscious patient is raped but prior to regaining consciousness the rapist kills the patient. In this case, the patient will not ever experience the harms of the rape but she is wronged *by* that rape.

Each of these cases is intuitively a case in which something is harmed but there is no interest that is violated or frustrated, at least not the interest of the thing that is harmed. Saving TRIA by explaining the harms as frustrating so-called ideal desires or interests is a non-starter. Ideal desires are one way of saying that there are objective goods the deprivation of which harms one. But it is the deprivation of the good not the unfulfilled interest that is the crux of the harm. It is also worth observing that cases (c) and (e) might function as direct counterexamples to the Experience condition, discussed above. Of course, at this inductive stage, these cases are not definitive. In fact, the inductive stage can get quite complicated.

Consider for contrast, the following three cases that seem to suggest the importance of experiencing harms.

(Timebomb): Suppose Jones attaches a timebomb to your bicycle with the intent of blowing you up. However, you crash and die before the timebomb goes off. The intuition is that Jones does not harm you by blowing you up for the simple reason that you don't blow up.

the harmful effects of his actions are never experienced by you, and so, he does not harm you.

(Torpid assassin): Suppose Jones is a professional hitman who is paid a modicum sum of money to kill an innocent Prince. Upon seeing the Prince and lining his sights rather sluggishly, Jones pulls the trigger and shoots a bullet at the Prince from a far distance. Before the bullet hits the Prince, the Prince falls dead of a pulmonary embolism. The intuition is that the Prince is not harmed by the assassin's act because he does not experience the effects of *that act*.

(SIDS): Suppose Jones abuses drugs while pregnant with Chuckie, but before Chuckie develops enough to experience the effects of that abuse, he dies of sudden infant death syndrome.

Some readers may have the intuition that Jones does not harm Chuckie in the SIDS case because Chuckie never experiences those harms. Others may, however, demur as I do. Not all harms must involve killing.

These cases illustrate that Jones gets morally lucky because the victims are not harmed by Jones's actions. And the reason is that they do not experience the harmful effects of *Jones's* actions.

How do our intuitions on these cases relate to our intuitions on cases (a)–(e)? I'm tempted to say not much at all. First, cases (a)–(e) are all cases in which an agent harms another agent. For the latter three cases, the victim dies accidentally, e.g., Jones *would* have harmed the Prince. Of course, one might claim[13] that the wrong-making feature in these latter three cases is a counterfactual condition: if the possible victim had not died first, that person would have *experienced* the harm of the bomb, the assassin's bullet, and so on. The victim's potential experience of the harm is the wrong-making feature.

This response, however, does not give evidence for the TRIA but assumes it and at best, provides a coherent story. At this stage of executing RE, however, we just have one explanation among several. Our moral intuitions can just as well be explained by the fact that Jones intended to kill an innocent person. Moreover, since the response admits counter-factuals, we can imagine a possible world in which the cyclist does not die accidentally, but has a suicidal thought right when Jones shoots him. Is Jones exculpated in just those worlds in which his victim has an interest in dying? At this stage of executing RE, we may log such a suggestion as counter-intuitive. At the very least, we cannot use our intuitions on the latter three cases and their counter-factual permutations as *favoring* TRIA.

Second, for each of the latter three cases, Jones' actions are still wrong. It is wrong even to attach a timebomb to someone's bicycle with the intent of blowing him or her up. It is wrong to fire a gun at an innocent person with the intent to kill that person. So, even though the putative victims are not harmed by Jones's actions (i.e., they are not killed *by* Jones's actions), there is still something wrong about Jones's actions. The explanation for such wrongdoing does not have to advert to any *experiences* of the victims or that their *interests* were frustrated by Jones. An

---

[13] I thank a reviewer for pressing this objection.

explanation for the wrong must account for the fact that Jones does something wrong and that wrongdoing is not merely to himself. He (somehow) wrongs the bicycle rider and the innocent Prince. Whatever the explanation is, what matters for my argument is that we have the considered intuition that Jones does something wrong and not merely to himself.

Third, the *structure* of cases (a)–(e) are different than (timebomb) and (torpid assassin). For the latter cases, there is a wrong action that is done and in one possible world scenario that action realizes its effects, namely, the timebomb blows the cyclist up and she dies. In the actual world, the timebomb does not go off because the cyclist dies in an accident beforehand. The timebomb *would* harm the person but does not. Because in both the actual and nearest possible world the victim dies, the victim's death is modally overdetermined, i.e., in both scenarios the cyclist dies. Cases of an elective abortion, or research ethics abuses, and so on, are not cases where death is modally overdetermined because in the nearest possible world the child *lives* or the research subjects are not harmed.[14]

Furthermore, cases of modal overdetermination are exculpating only in the scenario where the *accidental* death occurs—the death occurs without anyone's agency. But in the case of the child who dies from an abortion, however, the killing is a result of one's agency. The condition for exculpation is not met—unless the abortion is done on a stillborn fetus unbeknownst to the abortion doctor.

I should be clear that I am not assuming ~(pk) in drawing attention to the structural dissimilarities between the groups of cases. It is enough to point out that our moral intuitions on the modal overdetermination cases are such that Jones is exculpated only because death is overdetermined by a freak accident. Notice that if the cyclist is killed by a bigoted motorist, then the cyclist is wronged by *that* motorist. Jones is exculpated because death occurs by some other means. Again, that structural feature is not duplicated in the abortion case. It may be that abortion does not harm preconscious human beings, but that cannot be argued for *simply* by pointing out that the cyclist or the Prince do not experience any harms *by Jones's actions*.

3.5.2 *TRIA and Inconsistent Norms of Acceptance*

So, it looks like there are cases that fit into Class IV and therefore, there is inductive evidence that interest-violation and wrongdoing are not coextensive. Instead of building up to the TRIA by imagining cases, however, let's assume it. What follows if we provisionally accept it? Nichols explores this latter procedure. Given TRIA, it seems permissible to kill Alzheimer's patients and born infants since in both cases,

---

[14] The reader may ask what about fetal anomalies according to which the developing human being would die anyway? The point in the paragraph of course is to point out a structural feature of the different sets of cases. And that difference suggests that experiencing a harm is not necessary to be wronged. It is not that our intuitions conflict either; our intuitions are different on logically different cases.

there is only a present weak interest in living, if at all, and if they are killed, there will not ever be a future interest that is violated.

> Since killing a victim of Alzheimer's disease would not frustrate any of her future time-relative interests (because it prevents them from arising), and since her time-relative interest in continuing to live is so weak, the TRIA implies that it would be permissible to kill the Alzheimer's patient. This is implausible, and it is directly implied by McMahan's reasoning that abortion does not frustrate the fetus's future time-relative interests, because it prevents them from ever arising (Nichols, 2012, 502).

The same type of reasoning applies to infanticide. Importantly, both McMahan (2002, 342–346) and DeGrazia (2005, 289 ff.) grant that accepting TRIA would entail revising our moral intuitions about such cases. So, given TRIA, it is permissible to kill patients with advanced neurodegenerative diseases, and infants. Both cases of killing violate our considered intuitions as McMahan and DeGrazia agree, they just think that those judgments should be revised.

Notice what just happened. An adjustment decision is made, but as Haslett points out, what grounds adjusting one's axiological view in the way recommended by McMahan and others? One adjustment decision is to infer from our intuitions on prenatal impairment cases, cases (a)–(e), the Neuron removal case, the Alzheimer's and infanticide cases to a more inclusive view of wrongdoing—one that is more inclusive than a violation of interests. On the other hand, one can accept TRIA with its costs. But the latter decision is problematic since the TRIA rejects other axiological views because they fail to countenance one or more of our considered intuitions. Nichols explains the problem as follows:

> But the advocate of the TRIA now appears to be subjecting his theory to less stringent criteria of adequacy that those to which he subjected [other accounts of value]: after insisting that [other accounts of value] should be rejected because it fails to adequately accord with some of our case-based moral judgments, he claims that we need not reject the TRIA on the basis of a similar failure. This is an unacceptable methodology (Nichols, 2012, 502).

What follows is that TRIA is hardly the best there is. It inherits philosophical costs as well, and its claim to being the best is based on a prejudicial methodology.

Either there is a reason for making the latter adjustment decision or there is not. If there is not, we encounter an obvious de jure defect. What if there is a reason? What could that reason be? At this advanced stage of executing RE, why think that it is still the *best* account of wrongdoing? My suspicion is that the intuitive costs of accepting the TRIA are tolerated only because doing so retains a putative justification for (pk). The next section explores this suspicion and argues for Claim (iii) above.

3.5.3 TRIA, (pk) and Logical Circularity

The point in this section can be illustrated by seeing what happens if we assume (pk) at different levels of credences or confidence levels. Given the results of the previous two subsections, how should our belief in (pk) modulate in relation to the TRIA justification?

Suppose one believes (pk) to a credence of.5. (1.0 would be 100% certainty in its truth,.1 or 0 would amount to certain disbelief in (pk). A credence of.5 would be an attitude of something like 'it could be true, it could be false'.) Suppose also that one considers cases of impairment, putative wrongdoing without frustrations of one's interests as in cases (a)–(e) above, and neuron removal. Upon such considerations one might think that Class (IV) is non-empty and that pre-born human beings killed by abortion share intuitively similar features as those harmed in Class IV. Given the inductive evidence of our considered intuitions one has no reason for *increasing* her credence in (pk). That is because we have no reason for thinking that the TRIA explains *all* our considered intuitions on harm or wrongdoing.

One might be struck by the cases that suggest the importance of experiencing the harms. But I argued above that using the Experience condition as a reason to preserve the TRIA's justification for (pk) must assume (pk). Because on the present hypothesis one's credence in (pk) is.5, this riposte is not available—it would not increase one's credence in (pk). Likewise, the previous section noted that an adjustment decision had to be made and the latter decision is ungrounded. (And if it is grounded, we do not know what justifies it.) So, if one starts with a credence in (pk) to the tune of.5 and one executes RE, she has no reason for increasing her credence after collating her considered intuitions on cases of harm, wrongdoing, and their relationship with interests.

Suppose that one holds (pk) to a credence of 1.0 or.9 and considers the set of considered intuitions spanning Classes I–IV, in particular IV. If one already is certain in (pk) and recognizes that in order to preserve a coherent justification for (pk) one must hold the Experience condition, one's credence in *that* condition will go up. Likewise for the latter adjustment decision noted in the previous section. This is how anyone would think about it, including me. The point is that a deflector to an objection to the TRIA justification for (pk) can easily be accepted once one privileges (pk) in one's overall noetic commitments. RE tolerates making the adjustment decision in this direction. The Experience condition would then be accepted *because* (pk) is accepted, and this recapitulates the basic pattern of circularity I explained in the introduction.

The credence method delivers the following argument. Either (pk) is entertained with a credence of.5 or 1.0. If the former, we have no reason for raising our credence after canvassing the entire set of considered intuitions. If we begin the dialectic being certain about (pk), then propositions that make the TRIA justification coherent but have no independent justification, would be accepted *because (pk) is privileged in one's noetic commitments*. The former starting point tells us that TRIA fails *to justify*; the latter starting point tells us that the TRIA justification is logically circular. Both results entail that the TRIA justification fails to justify (pk).

Faced with these de jure defects in one's justification for (pk), the TRIA proponent should realize that at best, she might *be justified* in believing (pk), but not be able *to justify* (pk). She might execute RE impeccably, doing the necessary 'nipping and tucking' of her considered intuitions and axiological explanations for them. But doing so would at best, leave her with a coherent network of mutually supporting claims.

In the setting of peer disagreement, however, the proponent of the TRIA justification for (pk) has no epistemic standing.[15] She is like the sailor in the submarine who receives messages from other sources that conflict with what her own instruments are telling her. Recall that the sailor is not told whether she is in a real submarine doing a real exercise or in a simulator. In the setting of peer disagreement on moral matters, there is not a Archimedean perspective from which one can adjudicate the veracity of her faculties or her inquiry. She must take seriously the reports of others and treat their 'instruments' as possibly veridical. The TRIA proponent for the justification of (pk) is in a similar epistemic cul-de-sac and cannot get out. Peer disagreement is higher-order evidence that her 'instruments' may have rendered a processing error. And there is nothing within her coherent network of equilibria to deflect this higher-order evidence, not without vicious circularity.

3.6 The Bodily Rights Argument

There are two more arguments for the permissibility of abortion under strategy (II); under the assumption that what is killed is one of us. One of those arguments focuses on the bodily rights of the mother, and the second focuses on having the capacity for consciousness. Often enough, the capacity for consciousness is used as a criterion for personhood, other times the personhood of the developing human being is assumed but we don't have value until we have an exercisable capacity for consciousness. This chapter considers the latter iteration. This section discusses the argument from bodily rights. The seventh section discusses its de jure defects. The eighth section discusses the argument from consciousness and the limitations of focusing on that aspect of our worth.

The bodily rights arguments for the permissibility of abortion are important because they grant that the developing human being has a right to life, and yet, the conclusion of such arguments is that an abortion is permissible. Thomson's argument (1971) for instance, grants that the unborn child can have a right to life but not a right to the bodily support of the mother because the mother has bodily rights. The ingenuity of her argument is that it appears to grant everything a pro-life proponent wants, except a right to the bodily support of the mother.

If Thomson can motivate the view that a full-fledged person does not have a right to another's body, she will have motivated one aspect of her argument. She does this through the now famous Violinist case.

> You wake up in the morning and find yourself back to back in bed with an unconscious violinist. A famous unconscious violinist. He has been found to have a fatal kidney ailment, and the Society of Music Lovers has canvassed all the available medical records and found that you alone have the right blood type to help. They have…kidnapped you, and last night the violinist's circulatory system was plugged into yours, so that your kidneys can be used to extract poisons from his blood as well as your own. The director of the hospital now tells

---

[15] Chapter 5 defends the claim that there is in fact *peer* disagreement regarding (pk).

you, "…To unplug you would be to kill him. But never mind, it's only for nine months…"
Is it morally incumbent on you to accede to this situation? No doubt it would be very nice
of you if you did, a great kindness. But do you have to accede to it? (Thomson, 1971, 56).

Most will agree that there is no moral obligation to remain plugged into the violinist. (But most commentators also agree that it would not be impermissible to continue being plugged in.) Even if the Violinist has a right to life, the *scope* of this right does not include the use of another's body.

Boonin exploits this idea at length reflecting instead on an actual case, namely the *McFall v. Shimp* case (Boonin, 2019).[16] Robert McFall suffered from aplastic anemia, a disease of the bone marrow that leads to a life threatening diminution of blood cells. He required bone marrow transplantation from a compatible match. David Shimp was McFall's cousin and was in fact a compatible match. Shimp refused to donate his bone marrow even though doing so would not significantly compromise his own health. Though Shimp looks like a jerk, the court correctly ruled, in my opinion, that McFall does not have a right to Shimp's bone marrow. The analogies to the abortion case seem apparent: the developing human being does not have a right to access or use the mother's body for the child's sustenance and development.

Proponents of this strategy have faced difficulty in explaining how typical abortion procedures are analogous to disconnecting oneself from the violinist or Shimp refusing to submit to a bone marrow aspiration. Other issues pertain to the scope of parental duties, and whether there is a category of involuntary duties.[17]

3.7   De Jure Defects

3.7.1   Parental Duties, Involuntary Duties, and Consent

Warren (1973) critiqued Thomson's argument noting that our intuitions change if consent were involved. If, for example, you were part of a music lovers society the membership responsibilities of which involved taking a turn to be plugged into the violinist and your lottery number came up, you would have a duty to ensure the survival of the violinist. Warren's example is meant to parallel persons who engage in consensual intercourse. There is a well-known probability that pregnancy could

---

[16] Care should be exercised here since defenders of strategy II sometimes slip into comparing a woman's right to her own body versus the fetus' right to life—paired with an argument that the bodily right of the mother supersedes the right to life of the developing fetus. Strictly speaking, this is not how to understand Thomson-style arguments. How would one compare such rights? And if one were to do so, would not a right to *life* be more important than a temporary violation of a person's *bodily* right? It is better to understand the argument as limiting the scope of one's right to life. A right to life does not entail a right to another's body.

[17] On these and other points see especially Lu (2011), Finnis (1973), Beckwith (2007), Kaczor (2011), and Lee (2010).

occur with such a practice, and if one's 'number comes up' one is obligated to care. Other cases of this sort include gambling. One might engage in gambling knowing that one might lose money. The loss of money would be involuntary, but one must pay out if one loses. By engaging in the practice, one consents to the sequalae of that practice even if that sequelae are not desired.

Furthermore, there are cases in which a clear duty exists with no consent. Suppose your brother and his wife die in a freak accident and you are the closest family member to assume care for your niece and nephew. Arguably, you have an obligation to care for them. Your spouse is diagnosed with a serious disease that requires extended care on your part. Assume you do not like or desire the intensive care you must provide—e.g., caring for someone with early onset dementia. Arguably, you still have a duty to care for your spouse, at least for some time.[18] Cases like these suggest that our interest or desires are not reliable indices for our obligations.

There are other cases of consenting to an activity without consenting to the probable side effects of that activity that should be considered. Examples might include going out on the town at night. Getting mugged or having one's wallet stolen are not consented to. There is no reason to think that consenting to going out with friends at night entails consenting to being mugged or injured. Of course, there is a difference between these cases. It is part of the practice of gambling or being a member of the music lovers society that you pay up or take your turn caring. It is not part of the practice, if I can call it that for a moment, of going out on the town that one gets mugged. It is equally plausible to point out that it is constitutive of sexual intercourse that it is procreative. So, we have some cases in which involuntary side effects do not obviate one's obligations, and cases in which they do. And the cases for which they do not obviate our duties resemble consensual sexual intercourse and pregnancy.

Intuitions may also shift when we consider other disanalogies, one of which is that the Violinist is a stranger, but in an abortion, the 'stranger' is one's child. If it's the case that we have parental duties, duties we do not have towards strangers, then this is a disanalogy. But is it a relevant one? It matters what, forgive the pun, generates the parental duties. Brake's (2010) answer is that parental responsibilities are voluntary. "Voluntary acceptance of moral parental obligations is necessary, but not sufficient, for such obligations" (Brake, 2010, 152). I side with Prusak (2013) and several others, however, according to which the voluntarist account is implausible.[19]

To see why, let's grant a plausible starting point: Parents have special and unique duties to their children. And by parents I do not mean what many refer to as biological parenthood—Brake, for instance, includes sperm donation as an instance of

---

[18] I thank Michael Degnan for cases of this sort.

[19] Though I think the view is ultimately implausible, Brake's contribution is important and should anchor a broader analysis not undertaken here. She helpfully distinguishes different ways of viewing parenthood, legal, social, biological and so on. For instance, she includes sperm donation as an instance of biological parenthood. For the record, I do not, but she is welcome to carve up the landscape in that way. Also for the record, I do not see my view represented by causal accounts either.

biological parenthood. Because the discussion pertains to abortion, it is safe to circumscribe the scope of parenthood to a heterosexual couple who have engaged in consensual intercourse. Now why do they and no other person have such duties? No answer has a ring of truth to it except that those parents caused that child to be. Consider a born child or toddler. Let's imagine the parents taking care of the child, and let's also consider in contrast whether the next-door neighbor has any obligations to care for that child. Schwarz and Tacelli observe that in such a scenario the parents (including the mother) have "…certain special obligations. But why is that? Why does she, and not some other woman, have these obligations to him? There can be only one answer: Because she, and not some other woman, conceived him; because she is the biological mother" (Schwarz and Tacelli, 1989, 83). Their point of course is to envision a scenario in which we compare the obligations that the mother has to her child in contrast with any other woman vis-a-vis that same child. What grounds those unique obligations is that the mother was pregnant with that child.[20]

Consider some cases to illustrate the existence of such duties. Since bodily-rights arguments assume that the unborn human being has the same rights as a person, consider our duties to born children. Consider a burning schoolhouse scenario according to which you have the chance to save only one of two children. One child is yours the other is not. Whom do you choose? It seems clear that you choose yours—all else being equal. This intuition might not indicate a duty to save—for that we might need to know more details about the risks to you in saving one of the children; but it does indicate that if there is a duty to save in such a scenario, it is a parental duty.

Now suppose that there are two women and only one child is in the burning schoolhouse. Suppose that this time there are considerable risks in saving that one child. One woman is the mother of the child and the other woman is a stranger. It now looks like the mother has the obligation to save the child, the other woman does not. At the very least, there is a difference in moral evaluation if we consider some counterfactual scenarios.

Suppose the schoolhouse is on fire again and there are two women and one remaining child can be saved. In this circumstance suppose that there is very little risk to the women if one were to save the child. Suppose *both* women refuse to save. In such a scenario we view the mother with greater opprobrium than the stranger, indicating that the mother has greater responsibility for her child.

Moreover, we can easily modify these scenarios to illustrate that you still have a duty to a child simply because she is a living human being. Suppose the schoolhouse is on fire again and there is one child (not yours) who needs saving and you

---

[20] Adoption is not a counter-example to this claim since both legally and morally, the mother transfers those obligations to the adopting parents. In these respects, adoptive parents are parents (Brakman, 2014). Furthermore, it does not matter whether *the pregnancy* was wanted or not, though intentionally engaging in a *procreative act* is. See Prusak (2013) for more detail on why a causal account of parental responsibility explains our moral intuitions better than a voluntarist account even if the pregnancy is unintended.

are the only one who can save the child. Suppose that the risks to your life and health are low, but also that the child's parents died in the fire leaving the child orphaned. You should still save the child; and upon saving you and the community should ensure a good upbringing for the child.[21] This is how we treat orphaned children, migrants, and those seeking amnesty. I mention these latter cases to note that there is a *legal* framework that aims to accommodate them. Vulnerable unborn children are along the same continuum.

A voluntarist account does not explain our intuitions on these cases. It does not explain why *the parents* have such duties, and it fails to explain how the parents (before they procreated) *acquired* such duties. I did not acquire the duty to care for my child only when I consented to those duties. Consent may be required to act in accordance with those duties, since it is psychologically difficult to act according to norms with which one does not agree. But consent does not generate them.

Don Marquis (2010) extends these points observing that pregnant women are *mothers* to their unborn children. Since all mothers are parents and parents have special duties of care towards their children, pregnant women have special duties to care for their unborn children. If parenthood begins with pregnancy, then mothers have special duties to their unborn children as well. Does parenthood begin with pregnancy? On Strategy II, we may suppose that it does since on this hypothesis, it is granted that the developing human being has the same worth as a person.

In general, whether one has a duty towards Smith is *partly* a function of whether the person caused Smith to exist. Not all duties, of course, can be pinned on who caused Smith. But *that* Smith's procreators have duties to Smith is apparent. Suppose the parents decide against having a child but after the mother already gave birth. Do they relinquish their responsibilities for the child? It seems counterintuitive to say yes, as this would permit child neglect. It does not morally matter on strategy II whether this decision is made before or after birth since that strategy assumes the personhood of the developing human being. It would be quite surprising to find out that parents have the duty to care for their child by feeding, clothing, bathing, reading to the child, walking the child, and every other act of caring except gestating.

One might argue against these arguments noting that they all presuppose that the unborn child has a very specific right: a right to the mother's body. Cases like the violinist are meant to show that in some circumstances, persons do not have that right. Such a response is implausible as applied to the parent-child relationship. Whether the child's life depends on the parent's behavior or body does not morally matter (viz., behavior is embodied). If dependence upon another's body justifies neglect because it is not willed, so does dependence on one's behavior if it is not

---

[21] One might not share this intuition, but that is a dangerous undertaking. One need not share the intuition in Boonin's cases discussed below of the mother who is implanted with her child. The mother, one might say, consented to an IVF cycle, those are her children, and she and her spouse need to care for all their children. This sentiment does not strike me as crazy.

willed.[22] Imagine parents refusing to feed their baby because they do not want a baby. Such a position is not extolling the value of liberty, so much as it is tolerating child neglect.

Even if it is true that the *child* does not have a right or claim on the parent, *the parent* still has a duty to care. A person may have a duty to another even if the other has no right or claim on that person. McLachlan explores this suggestion and explains it as follows:

> Even if no one had a right to our bodies or a right to a particular performance from them, we might still have duties regarding our bodies. We might have a duty to keep our bodies healthy even if no one has a right that our bodies are healthy. A woman might 'own' a fetus totally in the sense that no one has a right that it is treated in any particular way and yet have duties regarding its disposal… (1977, 201).

"Disposal" might not be the correct term here in that McLachlan means welfare given the example that follows. He considers a case in which the human race borders on extinction and there is one pregnant woman in the entire world. We can see why she may "have a duty to have a baby rather than an abortion even though no one else had any rights regarding the treatment of her fetus" (1977, 201). This holds true even if the child has no right to be born. So, even if *the child* does not have a right to be gestated in his mother's womb, it does not follow that *the mother* does not have any duties to the child.[23]

One might argue, however, that the position that parents have special duties to their child yields counterintuitive results for the pro-life position. At issue is whether the disanalogy between stranger and child is morally relevant. Suppose, as I have argued, that it is. Boonin (2019, 90 and 92) however, has us consider several forced implantation cases according to which a woman has an embryo forcibly implanted in her uterus. Suppose in one scenario the embryo is hers from a previous IVF cycle; in another, suppose it is not hers. Fill in the narrative as you like (e.g., she is kidnapped by an otherwise well-meaning infertile couple who happen to own an IVF clinic). Two points follow: if the stranger-child disanalogy is relevant, then it looks like the pro-life position should permit that the woman with the embryo that is not hers may abort. That child is analogous to the stranger in the violinist case. If you are permitted to unhook yourself from the violinist, you may 'unhook' yourself

---

[22] Some (Boonin, 2019) think that duties to others are generated because one harms the other. If I deliberately inoculated you with the flu, I'm obligated to go to the drugstore and buy you medicines to moderate your symptoms. Let's assume that all cases in which one agent harms another generates a duty on the wrongdoer to repair that harm. This assumption entails nothing about one's duties in cases of creating another. All Dobermans are dogs; but all labradors are also dogs. So too, all cases of harming may generate duties to care. But this is compatible with the claim that all cases of creating generate duties to care as well.

[23] McLachlan offers another example to illustrate the point. Consider an apocalyptic scenario in which the world and its resources are paucious, but you have more than enough food, water, and resources for the present and future. Presumably, you have a duty to share your resources, maybe not a duty to *everyone*, but a duty to share to others (on some just tirage principle). Any given individual who comes to you does not have a claim on your food, but you still have a duty to share. So, a duty to care can be decoupled from whether the cared for individual has a right to that care.

from the child who is not yours. Second, it also seems counter-intuitive (for Boonin and maybe some others) to say that the woman with the embryo that is hers should gestate. The implantation was not consented to. That fact alone is enough to undercut any duty to gestate even if it is her child.

Several observations are worth noting before addressing the specific argument Boonin is constructing here. First, he is formulating cases that are morally analogous to cases of rape—i.e., forced implantation. Second, he is more concerned with what the state would be justified in doing, not whether the actions in question are morally permissible or impermissible. I glean this latter point because he is, as noted above, testing our intuitions against the actual *McCall v. Shimp* case. Third, Boonin grants (2019, 89), that if we revise the *Shimp* case to parallel a mother-child relationship, a similar refusal to save the life of her born child through a similar donation procedure makes the mother look morally suspect—"we have special legal obligations to our own children" (2019, 89). I think Boonin is right on this point. But it is a concession that circumscribes even further his argument. Suppose that the mother refuses to donate her bone marrow to save the life of her child because the child is getting in the way of her career advancement or education. Most of us should think that she should still donate.[24] Again, we do not call this liberty, we call it child neglect.

Fourth, after refusing, what Shimp did not do is dismember and intentionally kill McCall. But, such actions would be more analogous to most abortion procedures.

Boonin's conclusion is that there should be no state imposed restrictions on procuring an abortion, viz., a hysterotomy given what Boonin says at (2003, 193) and the discussion in Sect. 3.7.2. And the circumstances are specific to cases of rape (viz., forced implantation).[25] In such scenarios, there is no duty—parental or otherwise. This is quite a narrow conclusion derived from highly artificial cases. I provided at the beginning of this subsection numerous cases illustrating that we have a duty to care that is not a function of one's consent or dissent. In the setting of the current stage in the dialectic, Boonin's case is quite weak. There is not a strong

---

[24] In setting up the example this way, I am setting aside a worthwhile discussion (and critique) of social policies that make motherhood difficult, or the lack of responsibility exercised by many male sexual partners.

[25] There are a few notable observations about using cases of rape to advance one's argument for (pk). I find it (1) a distraction to focus on the issue. Proponents of (pk) do not support restrictive abortion laws that leave exceptions for rape. There is a different reason offered in favor of (pk). (2) It is hard to see how the circumstances of our conception affect our worth. Serrin Foster (personal communication) relates the following thought experiment. Imagine four of your best friends. Imagine finding out that one of your best friends was conceived in rape. Does that lessen your evaluation of that friend? Clearly it shouldn't. The circumstances of our conception do not affect our worth. (3) Finally, many say that not having an abortion will remind the victim of her assault and traumatize her over again. There is no doubt that trauma in this setting should be treated or minimized. But considered as a reason for (pk) it is weak. On this reasoning, a woman who suffers an acrimonious divorce may kill or never see her children since they remind her of her vicious ex-husband and the trauma of dealing with him. Alternatively, one could just as well see the child as a reminder of the mother's own heroic act to care for the child. Nevertheless, none of these observations argue for ~(pk), and I find myself having similar intuitions as Boonin on his cases.

argument here in favor of no duty to gestate, particularly when we consider cases of saving born children, individuals seeking amnesty, and displaced migrants (see above). We make sacrifices to care for and ensure the good of innocent and vulnerable human beings. The same can be said for unborn human beings. Gestation can, after all, be understood as a form of caring. Nothing Boonin or others say force us to understand it as a morally neutral act.

But how should we assess the mother who is forcibly implanted with a child not her own? Does she have a legally enforceable duty to gestate? And if we say no, what follows for typical abortion cases? Again, if the pro-lifer says that the stranger-child disanalogy is relevant, it looks like it follows that the 'stranger' (i.e., this child that is not hers) may be killed.

If the bodily rights argument is meant to be an argument for (pk), it must say that no reason for impermissible killing is present in the imagined cases, and a premise to the effect that the target case is sufficiently like the imagined cases. Being one's child is one reason for not killing, *as is being an innocent and vulnerable human being*. It seems perfectly consistent to say that one may unplug oneself from the violinist, but not 'unplug' one's child in a typical abortion scenario because of a parental duty. But our moral obligations are not exhausted by parental duties, as indicated in the cases canvassed above. Simply put: we have duties not to kill children, whether our own or not.

But doesn't this response apply to the stranger too? Two observations are noteworthy. First, think of what is happening to our intuitions as we execute RE. If we consider first the cases canvassed above indicating a duty not to kill one's child (viz., schoolhouse scenarios, involuntary duties, and McLachlan's cases) and then consider the violinist case, one may find it counter-intuitive but on pain of inconsistency, accept that one needs to keep the violinist alive. Conversely, if one considers first the violinist case and then considers the cases canvassed above one may find it counter-intuitive but on pain of inconsistency accept that the children in such scenarios may be allowed to die. This is the best a proponent of the bodily rights argument can hope for given a fair assessment of the points noted so far. In such a setting, the bodily rights argument is weak and unpersuasive.

My second observation is more damaging. I don't think the dialectic ends with such a stalemate. There are several cases that mirror the violinist or forced implantation cases in relevant ways but our intuitions suggest saving even at immense personal cost. Hershenov (2001) entertains several cases of this sort. Consider a marine biologist doing work in a remote part of the northern Pacific. He is on a raft with necessities such as food and water but also his equipment which includes expensive gauges, radar, and electronic data storage that includes all the results of his studies out at sea. The data he has collected, we may assume, represents the culmination of his life's work and we may suppose it is unrepeatable given the rare species he has been studying. He has arranged for a pickup after several months of work. But before he is picked up a cruise ship passes by and, as Hershenov describes, the ship explodes. Neither the biologist nor the pilots of the ship are able to send an SOS. Everyone is killed "except for one small child who will soon succumb to the frigid waters if not pulled from the sea. But there is no room on the raft for the child unless all of the irreplaceable expensive equipment and data are thrown overboard and forever lost" (Hershenov, 2001, 133). In such a scenario the biologist must save the child even if it means discarding his life's work overboard. We may even suppose that in accommodating the child on the raft one must sleep in an uncomfortable position until the rescue boat arrives.

Here's a case in which a person is uniquely positioned to save another. But saving the person involves considerable sacrifice and bodily discomfort—features of which parallel both violinist-type cases and pregnancy. But our intuitions are on the side of saving, not letting die.

Consider another case. Suppose two birdwatchers are on a remote island but are taken up by a sadistic booby-trap that nets them above a crevice. At a certain point the net will be released, letting the two fall into this crevice. On one scenario the larger person knows that if they fall together he can shield the smaller person from being killed by the fall. "The larger person knows that due to the way he will hit the ground, he will suffer nine months of intermittent back pain, nausea, and abdominal swelling comparable to what a pregnant woman bears" (2001, 134). On the other scenario the larger person can have a third party cut only him away from the net leaving the smaller person to fall; and because the way the smaller person is positioned, the latter would suffer fatal injuries on impact.

> The larger person is a stand-in for the pregnant woman pondering the abortion of a fetus that is considered a person, as well as being in a situation similar to that of a person who has the violinist attached to his healthy kidneys. The third party on the platform who helps the larger person detach himself is equivalent to the medical practitioner who performs abortions or facilitates the disconnection of the violinist. The smaller person entangled in the ropes is in a position comparable to that of the violinist as well as the fetus if we assume, as Thomson tells us to do, that the fetus is a person (Hershenov, 2001, 134).

Intuitively, the larger person should stay in the netting and save the smaller one from a fatal impact. Hershenov observes that even if we understand why the larger person would want to be cut free, it still would be wrong not to save.

Though it might be fun, momentarily, to play thought experiment Olympics—designing the most imaginative and artificial cases—there are important philosophical points to make. There are cases that are analogous to violinist-type cases and for which we have the intuition to save. The existence of such cases reveals that our moral intuitions are quite labile—they are latching onto subtle and possibly morally irrelevant features. Suppose, for instance, that kidney failure—which is what affects the violinist—was as widespread as the dependency relation in pregnancy. I would say that providing support is required (Hershenov, 2001). Those settings in which dependency is a norm, are precisely those in which caring is also expected. It might be hard to imagine widespread renal failure, so instead consider our elderly parents who progress to further stages of dependency. That progression is the norm. But no one questions the duty to care for one's elderly parents. That care is written into our practices as adult children (Brakman, 1994). Caring and dependency are correlative. So, our intuition that we may free ourselves from the violinist might be a function of the sheer unusual nature of the case. Considering cases of dependency closer to home, so to speak, reveals that it is not at all counter-intuitive to care.

Another observation about the reliance on our intuitions on these cases is to pay attention to their implicit assumptions (Napier, 2016). There is something worrisome about Boonin's cases in particular (2019) which involve forced artificial insemination or implantation. We are to assume as morally neutral the apparatus of IVF and the practice of freezing human embryos. It seems right that it is hard to spot the *parental* duties in such scenarios—or any duty. One can imagine various permutations of such cases the point of which is to illustrate that parental duties, qua pro-creator, do not exist (e.g., sperm donor). The cases for which we find it hard to spot who has the duty to care for the child, however, are precisely the cases for which we should view *those* actions with suspicion. If the way in which children are created is by actions that make it hard to see why we should care for them, that is a moral problem with those actions. With these thoughts in mind, one may coherently hold that there is no duty to gestate *and* it is still immoral to kill the child. This is coherent if one thinks two other claims: an immoral action was done that undercuts the *parental* duty (there may be other sources of our duty to care); and one immoral action does not justify another. Even if we agree with Boonin that there is no duty to gestate in a certain circumstance C, it does not follow that one may permissibly kill in C.[26]

3.7.2 Kaczor's Equality Argument

So, duties to care or save exist in many circumstances where consent is irrelevant. Furthermore, there are intuitive suggestions to the effect that parenthood entails having unique duties and these duties start when one becomes a parent. Given strategy (II), one becomes a parent by causing a child to be. Our intuitions on *when* these duties exist do not comport with a voluntarist account as indicated by entertaining cases of born children.

The argument in this subsection accepts the assumptions of strategy (II) and it grants that you have a right to unplug yourself from the violinist. You do not, however, have a right to dismember and kill him. In the best scenario, the Violinist case

---

[26] One last comment is worth observing on Boonin (2019). Again, he is discussing biological parenthood, and whether such parenthood generates a duty to gestate. To quote at length "If the reason you think it would be wrong for the state to let Alice have an abortion is that Al [the developing human being] has the right to use Alice's body because Al is Alice's biological son, then you'll have to agree the state should let Allison [a different woman] have an abortion because Alvin isn't Allison's biological son and so doesn't have the right to use Allison's body" (Boonin, 2019, 92). Unfortunately this is a formal fallacy. To see this consider the first conditional according to which if a woman W is the mother of her son S, W should care for S–W may not abort S. The second half of the conditional supposes that Alvin *is not* Allison's biological son. So, if it is not the case that W is the mother of S, nothing follows. If I am in Chicago then I am in Illinois. If I am not in Chicago, it doesn't necessarily follow that I'm in Illinois. This is the well-known fallacy of denying the antecedent.

argues that a right to life does not entail a right to someone's body. It does not follow from this that typical abortion procedures are permissible. Why is this the case?

If the mother has bodily rights because she is a human being, and the unborn child is also a human being, it follows that the unborn child has bodily rights as well (Kaczor, 2011, 151 ff.). The unborn child has a body and he is functioning and developing as he should. If human beings have bodily rights, and the unborn child is a human being, he has bodily rights too. What is the content and scope of these rights? Kaczor correctly observes that they include not being killed! Human beings have bodily rights because all human beings are *embodied.* Since even unborn human beings have bodies, it would come as a shock to find out that they did not have *bodily* rights. It would also come as a shock to find out that of the bodily rights we do *not* have is a right not to suffer bodily dismemberment.[27] So, with almost identical premises as the bodily rights argument, we can derive a reason for ~(pk).

The principal objection to this bodily rights argument (for ~(pk)) is that it only applies to those abortion procedures that dismember or directly attack the unborn (e.g., D&E abortions). The argument does not apply to procedures that remove the unborn from the mother's womb (e.g., hysterotomy). Boonin, sensing the need for this rejoinder, explains it as follows. "[I]n both of these cases [letting die and hysterotomy], abortion seems simply to be a means by which a woman who has been providing needed life-support to the fetus she is carrying can effectively discontinue her provision of such support" (Boonin, 2003, 193).

Notice what this circumscription must grant. The response is that removing the unborn human being from the mother's womb so that the unborn child dies, is compatible with the child's *bodily right* not to be intentionally dismembered.

To circumscribe bodily rights so narrowly makes it implausible that pregnancy could count as a violation of the mother's bodily right too. How so? We are asked to believe the following claim: Intentionally letting a healthy unborn baby die, as with a hysterotomy, is compatible with respecting the bodily rights of that baby. Presumably, this is because the baby is not directly attacked or dismembered. Well, so too with pregnancy. The pregnant woman is not attacked or dismembered by the growing child in her womb (Eberl, 2010). A healthy pregnancy, after all, is defined as *both* the mother and child being healthy (Hershenov, 2020). The pregnancy might be unwanted, but that would not entail that it is an attack, much less a dismemberment. So, if a hysterotomy is compatible with respecting the baby's bodily rights— because it is not an attack or dismemberment—a pregnancy is not a violation of the mother's bodily rights either because it too is not an attack or dismemberment.[28]

---

[27] McLachlan (1977) thinks that only persons can have a bodily right. My arm, or leg, or kidney, does not have any rights; only *we* do. This idea sounds right but, on the assumptions granted by Thomson and all who follow strategy II, it is a banal observation. The developing human being is assumed to be one of us, so its body has rights. It is also a banal observation if persons are embodied beings.

[28] For those who are still concerned about how these reflections handle unwanted pregnancies, return to the original claim that a hysterotomy done *so that* the unborn child dies is *not* an attack on that child. What must be meant by the term attack here is that there is no causally direct action

One might say that the relevant bodily right is the right not to have one's body *used* by another. In response, if one has a bodily right not to have that body used by another, it does not follow that the unborn child may be intentionally killed even by means of removal. The reason why is that born children need their parents for their survival as well. They need their parents to behave in such a way that is supportive of their growth and maturation. In this sense the children have a claim on the bodies of their parents (insofar as behaviors are embodied). The functional effect of a hysterotomy is to deprive the child of nourishment and oxygen. But parents are obliged to ensure that their born children are fed and have access to oxygen. That right is violated when nourishment and oxygen are intentionally deprived when they could be maintained by following the natural course of a healthy pregnancy. It is important for understanding this rejoinder to focus on that fact that Thomson and Boonin grant the personhood of the unborn human being. Therefore, if the use of the parental bodies is morally equivalent between gestation prior to birth and typical care after birth, the right not to have one's body used by another does not justify the intentional killing of one's child—even if by removal.

Consider Schwarz's (1990) stowaway case to illustrate these points. He entertains a case in which a stowaway on a ship is found half-way through the boat's journey.[29] The captain has only two options: throw the person overboard or keep him on the ship until port (there is no third option of not supporting or helping him). Throwing him overboard is a clear violation of his bodily right not to be intentionally displaced into an environment in which his body was not meant to survive— namely, water.

Schwarz's stowaway example is meant to mirror unwanted pregnancy: one has a person 'onboard' and desires that the stowaway not be there. Throwing him overboard is unlike withholding sustenance, but much more like intentional killing via displacement. Likewise, hysterotomy (prior to viability) for an unwanted pregnancy looks more like intentional killing via displacement than withholding sustenance. Brahms (2019) observes several other examples to illustrate the same point: if one is scuba diving with another and removes the other's mask and he drowns, that is less direct than knifing him in the thoracic area. But an action that is *less* casually direct is still an intentional killing. Stoning someone to death is less causally direct than beating him with one's own fists. But both are intentional killings. Likewise, intentionally depriving someone (however old) of oxygen and sustenance through either hysterotomy (or chemical abortion) is killing, not refusing to support.

---

on the child's body rendering it incapable of living. These features are not present in a healthy pregnancy—whether wanted or not. Consider what we would say about a wanted pregnancy. They are not considered direct attacks on the woman's body. But the property of wanting a pregnancy is not a bodily property anyway. So, a pregnancy per se is not an attack, but a sign that the reproductive system is functioning as it should.

[29] I follow Timothy Brahm's excellent discussion of this case in what follows. See Brahms (2019), "Blood donation and bodily rights arguments," available at https://www.youtube.com/watch?v=YmBrUcpOxDw (accessed 21 June 2024).

So, the conditions under which our bodily rights are violated are broader than direct attacks—as the previous points illustrate. Let's collate the results of the past several points. This dialectic started with Boonin's retreat to the claim that a hysterotomy is justified by a Thomson-style argument. We discovered, however, that if a hysterotomy respects the baby's bodily rights, it looks like a pregnancy would not violate the mother's bodily rights—even if it is unwanted. Schwarz and Brahms suggest that our bodily rights may be more inclusive than Boonin suggests. We have a bodily right not to be displaced into environments where we would die—viz., being delivered prior to viability so that the child suffocates.

But these results invite the following argument. Either a violation of one's bodily right is understood narrowly as protecting one against direct attacks *alone* or they are understood more inclusively to protect against intentional killings involving varying degrees of indirectness such as displacement or deprivation. On the narrow understanding, a hysterotomy does not violate the developing child's bodily rights. But a mother's bodily right is not violated by a pregnancy either since a pregnancy does not involve a direct attack on her bodily functioning. This means that the hysterotomy reply is not available to the proponent of the Thomson-style argument— i.e., the mother's bodily rights are not violated by a pregnancy (wanted or not) since the child is not directly attacking the woman's body. Given the inclusive understanding, the baby's bodily right is violated by procedures like a hysterotomy. On this inclusive understanding we have the right not to be intentionally deprived of sustenance. The baby too has a right not to be deprived of nourishment and oxygen. But a hysterotomy would be intentionally displacing the baby into an unsupportive environment and deprived of sustenance. On either disjunct, there is no argument for (pk).[30] There is not a de jure defect here so much as there is simply not an argument for (pk).

3.8 The Value of Consciousness

If we focus on the argument from consciousness, we can appreciate that it fares better than the previous argument. The central point of this argument is that we, you and I, do not become valuable or acquire a right to life until we develop an exercisable capacity for consciousness. No one can doubt that being conscious is a value

---

[30] It should also be noted that proponents of the bodily rights argument seem content to compare different rights: the mother's putative bodily right not to be pregnant (by any means) vs. the unborn child's right not to be intentionally dismembered and killed. These are different rights with different content (Gray, 2012). In that regard, they cannot conflict. For typical pregnancies, what are *not* in conflict is the mother's right not to be killed and the unborn baby's right not to be killed. There may be a conflict of rights, but not between the same ones. When rights-with-different-content conflict, we prefer whichever right protects the greater value. I do not know of any abortion proponent who has argued that not-being-pregnant is more valuable than sustaining the life of one's healthy child.

(Dung, 2024), nor that having the capacity for consciousness is a value. Neil Levy summarizes both the ubiquity and importance of this view on consciousness.

> It is widely assumed that consciousness matters. It matters in a way, and to a degree, that many of the other topics on which philosophers outside the explicitly normative areas of the discipline spend their time and energy do not. It matters in the kind of way that central topics in ethics and political philosophy matter. Indeed, it is, or should, be a central topic in ethics, since it is intrinsically valuable and/or confers value on those who possess it. Accordingly, consciousness, or its lack, is often invoked in discussions in applied and normative ethics…If the fetus is conscious, it is widely held, its moral value is greater than if it is not yet conscious, and this extra value must be taken into account in discussions of the permissibility of abortion (Levy, 2024, 127).

So, the view here is ubiquitous, and its implications for abortion and related issues are apparent.

There are plenty of considerations in support of this view. Consider the following thought experiments. Wellman has us imagine a world,

> not unlike ours, which contains no sentient beings at all. It might contain towering mountains, plunging waterfalls, moonlit pools, and sunsets; but it must contain no angels, people, or even animals. Now imagine that a single sentient mind enters this world to become aware of its rainbows and bubbling brooks. If changing this single variable introduces value into this world, it looks like consciousness is a value but things, no matter how beautiful, are not valuable in themselves (Wellman, 1971, 35).

Consider another thought experiment this time from Roland Puccetti (1983, 169).

> Suppose that one day a genie appears before you and convincingly displays magical powers. Now he tells you that he will, if you want, expand your brain so that you will have an IQ of 400. This means that whatever you're interested in achieving will come within easy grasp, whether it be leadership of state, heading a conglomerate, outpainting Picasso or getting a Nobel Prize in medicine. However, he explains, there is one hitch. If you want him to do this for you, you will at the moment of brain expansion cease forever to have conscious experience. No one will know this, for he will program your brain to make all the correct responses to questions, etc. Nevertheless the great future achiever will be an automaton and no more. Would you accept this offer?

Before I consider the limitations of these suggestions, I harbor suspicions about both. My questions for Puccetti's thought experiment are how can one not be conscious but have an IQ? How can one *intend* to win a Nobel Prize without being conscious that such an achievement is a good thing? I'll assume such an explanation can be given, maybe along the lines of contemporary developments in AI, but even these developments give me little confidence that it can *understand* winning the Noble Prize *as good*. Nevertheless, I assume that what the Genie can do here is possible.

Wellman's thought experiment may suggest only that consciousness of beauty is a value not that the world has no value *until* consciousness enters. Why? If consciousness were the only value, then the experience of the beautiful bubbling brooks and majestic mountains would not be a value separate from simply being conscious. But it clearly is. The thought experiment must presuppose that the world sans consciousness is still good or valuable. But until consciousness enters, so to speak, there is no *appreciation* for its value. That there is no appreciation of its value does not entail that it doesn't have value prior to that appreciation. I'll suppose, however, that requisite modifications can be made to motivate more interesting axiological claims.

As indicated, Wellman's and Puccetti's reflections are still opaque about which conclusion or claim they wish to motivate. Here are some suggestions:

1. Consciousness is the only value.
2. Consciousness is a necessary condition for a thing to have value.
3. Consciousness is a necessary condition for experiencing value.
4. Consciousness is a sufficient condition for value.

Consider some corollaries:

5. Things that develop (or have) consciousness have value.
6. Existing (as a human) is a necessary condition for having value.
7. Existing (as a human) is a necessary condition for experiencing value.

We can see how each option is quite different than the others and that it is unclear given the thought experiments precisely which claim is being motivated. Furthermore, some of these claims, upon inspection, do not require any motivation because they are tautologous, e.g., 3. That consciousness is a necessary condition for experiencing value is tautological if we understand by consciousness something like awareness. (But even this simple point is subject to qualification as some experiences don't involve an awareness as with states involving FLOW (Csikszentmihalyi et al., 2014). Wellman's thought experiment suggests 3, but charity demands interpreting it otherwise. Other claims do not require motivation because they are clearly false as with 1. Numerous things are also valuable, such as virtues, works of art and so on. As I see it, both thought experiments want to motivate 2 as against one of the competitors 5, 6, or 7. For simplicity, I understand the pro-life view as committed to 5. Do these thought experiments motivate 2 and disconfirm 5?

Consider Puccetti's thought experiment in reverse by which I mean that instead of stripping you of consciousness, the Genie's promises can be realized only by agreeing to be transported back in developmental time. Suppose that the Genie appears and offers to expand your brain so that you will have an IQ of 400, and so on. The only catch is that if you elect to be supercharged you would start your life back immediately after conception—assume all biological laws of human development hold as in the actual world. The other option is remaining the way you are now with no intervention. You might be undecided, but at the very least, you would not be irrational in choosing the supercharged world—choosing that world comports with your egoistic concern. If you choose the second option, you are still choosing to exist. So, viewed in reverse, the same thought experiment suggests instead that existing as a thing that *develops consciousness* is a necessary condition for value (i.e., 5).

As discussed in Chap. 2, if we have the same thought experiment—or two trivially different thought experiments—but two different conclusions can be drawn from them, they cannot justify only one conclusion. This is an instance of a justification-defect satisfying condition (i). Of course, if you think that my permutation is a different thought experiment, we still have an equally plausible thought experiment/intuition, and yet two different conclusions follow. This is a justification-defect satisfying condition (ii).

We can generate permutations of the supercharged world by extracting the feature of envisioning your existence prior to birth. Suppose you had a choice between two possible worlds becoming actualized. In the first possible world you don't exist at all. In the second you exist in the world as a healthy developing human being in your mother's womb and your parents have no plans to abort. If actualizing only one of these worlds is up to you, which world do you choose? You should choose the second. If you cannot decide yet, suppose your life in the second would be all things good, and you would die a painless death. Still undecided? Imagine that the second world has you experiencing joy upon joy with no hardship, challenge, or sadness, and the other world is devoid of your existence. It is clear that you would choose the second world in which you enter as a conceptus in your mother's womb. In choosing that world, you are choosing to exist; *you* enter it. By actualizing the second world, your choice comports with egoistic concern, you are choosing to exist in it. The fact that there is such a great difference between the first and second worlds suggests that in choosing the second you are not merely choosing *potential* existence; you are not choosing a world in which you do not exist. As with the previous thought experiments (i.e., Wellman's) one's choices are indicative of what is valuable. It follows here too that what you choose is indicative of the value of existing. Your choice also suggests that you are choosing your existence even though you are choosing to enter the world at a developmental stage in which you are still in your mother's womb. Because the only difference between these worlds is your existence at an incipient stage of development, in choosing the second world, it follows that a thing that develops consciousness has value.

Per the discussion of Chap. 2, it looks like we have iterations of justification-defects satisfying either condition (i) or (ii), depending on how one may individuate thought experiments. It is important to note that all the developmental permutations canvassed above used rational egoistic concern as a heuristic. And the results of these permutations confirmed 5, viz. *things* that develop consciousness have value. We do not, therefore, have an exclusive justification for 2 that does not fall prey to a justification-defect.

3.9 Conclusion

The problem I highlight in this chapter is that the TRIA justification for (pk) lands one in an epistemic cul-de-sac, specifically, an inference defect, an inquiry defect, and a justification defect (which satisfies either condition (i) or (ii)). To take just one defect, if the TRIA justification for (pk) is viciously circular, it fails to justify (pk). I do not argue that the balance of relative plausibility between the TRIA and other axiological theories is on par. In this regard, I do not consider *"tu quoque"* arguments, but even so, I find the conclusion that the TRIA is merely coherent and on par with alternatives still a significant achievement.

As I see it, the only option to controvert my argument is to explain why circular arguments can still function as good justifications for controversial moral claims. Of course, another option is to abandon the TRIA justification for (pk). In the next chapter we explore alternative justifications for (pk).

Chapter 4
Psychological Accounts of the Person and the Defense of (pk)

4.1 Introduction

In the previous chapter, I canvassed the time-relative interest account of harm or wrongdoing (TRIA) and explained how the TRIA functions as a premise in an argument for the permissibility of abortion (permissible killing = pk). Even if one thinks that you and I come into existence sometime shortly after conception,[1] when most abortions are performed, those abortions are permissible if the TRIA is correct. I argued that the TRIA justification for (pk) is viciously circular as well as other de jure defects. This circularity is brought into relief when we consider putative counter-examples to the TRIA, namely cases in which harm occurs without a frustration of one's time-relative interests, e.g., prenatal impairment. Deflecting such counter-examples required appealing to (pk) itself. Therefore, the TRIA justification for abortion has a premise in an argument *for* (pk) that is defended by *assuming* (pk). And that pattern of argumentation satisfies criteria for a viciously circular argument. I addressed another popular argument for (pk) that also assumes the humanity or personhood of the unborn human being, namely bodily rights arguments and the value of consciousness. The defects here are not best understood as circular, but rather weak justifications satisfying either condition (i) or (ii) (viz., Kaczor's equality argument).

In this chapter, the discussion shifts to consider the no-person strategy, labelled strategy (I) in the previous chapter. The no-person strategy strikes me as a much more plausible way of arguing for (pk) than any of the arguments discussed under strategy (II). The defense of (pk) rests on a successful argument that what is killed is not one of us.[2]

---

[1] For DeGrazia (2005) you and I come into existence after the possibility for twinning has passed. Boonin (2003) and Thomson (1971) hold similar positions.

[2] David Hershenov has pointed out a corollary for the pro-life position, namely, that a viable pro-life argument, excuse the pun, does *not* require arguing that what is killed is a person (Hershenov,

S. Napier, *Justified Killing*, https://doi.org/10.1007/978-3-032-14946-6_4

In bare outline, the no-person strategy aims to argue that you and I do not come into existence until well after conception. If correct, what is killed is not a person, by which I mean, the kind of thing reading this book; a rights-bearer, at least a bearer of rights apposite to human persons. If something is not a person, you cannot have the rights of a person. Otherwise stated, these arguments must argue that there are no *victims* of abortion because what is killed is not a person. The term 'person' in this context refers to the kinds of entities for which it would be wrong to intentionally kill. Most accounts of persons in such arguments conclude that persons are those human beings who evince exercisable psychological functions, such as consciousness, sentience, reasoning etc. In his typical clarity, McMahan states the view as follows: "An early abortion does not kill anyone; it merely prevents someone from coming into existence. In this respect, it is relevantly like contraception and wholly unlike the killing of a person. For there is … no one there to be killed" (McMahan, 2007, 186). I call this the functional view of persons because it emphasizes that to be a person with moral status the entity must have functional psychological capacities. McMahan's argument gives voice to what I refer to as the function argument. This is the no-person strategy. I begin with a section that explains the importance of the no-person strategy. I then turn to a detailed outline of how it defends (pk). The remaining sections canvas various de jure defects of the argument, principally the problem of underdetermination (Sects. 4.4 and 4.6).

4.2 Motivating the No-Person Strategy

The no-person strategy looks very promising but to appreciate its full utility we may consider arguments to the effect that it is inert. First, there are some cases in which one can wrong non-persons. Wreen explains,

> An unorthodox Kantian might say as much about a person who doesn't develop his talents, an environmentalist about someone who trashes the environment, an animal liberationist about an individual or society that allows a species to die off, or a natural law theorist about a couple that deliberately fails to reproduce-consider, for example, the last couple on Earth, after a nuclear catastrophe (Wreen, 2004, 560–561).

Wreen's point is that wronging can be decoupled from there being an individual who is wronged. McLachlan (1977), quoted previously, supposes that we may have duties to those who do not exist. If I promise a friend to harbour a secret even after he dies, I have a duty not to disclose (assuming there is no other conflicting duty). We have duties to follow the will of a deceased person. These duties, one might say, are a function of friendship or contract, but there are other cases of a duty existing without either of these. McLachlan explains, "I never met Wittgenstein. Yet, I would suggest that I have duties towards the dead man. For example, I have a duty not to

---

2019). That claim is compatible with the claim in the text, viz., a viable argument for (pk) requires arguing that what is killed is not a person.

slander him and it is not an obvious abuse of language to say that he has a right not
to be slandered" (1977, 199).

We may have duties without there being a corresponding right holder. Again,
McLachlan explains.

> Suppose that we were living in a country where many people around us were starving and
> that we had far more food than we needed for both our present and our future purposes. We
> would surely have a duty to give some of our food away, in the sense that we ought to do so,
> even if it were the case that no particular person had a right to receive food from us
> (McLachlan, 1977, 199-200).

It can even be true that the loss of one's rights can *initiate* duties toward that person.
A clinically insane person loses the right not to be coerced, or a right to make decisions for oneself and so on. The loss of those rights does not discharge the moral
duties we have towards that person. In fact, we might acquire *added* duties to care
for the insane person particularly in regard to the capacities he has lost.

Someone might lose certain rights but retain *some* others. But it might also be the
case that we have a duty to care for another thing, whatever that thing is, not because
the thing has a right to our care, or that it has a "claim" on our will somehow; but
just because it would be good for the thing to receive our care. And it is good for
living things, qua living, to receive care that contributes to their growth and maturation (Hershenov and Hershenov, 2017, Sect. 4.2). So, though it is commonly
assumed that one can wrong only persons, or that moral duties to another are parasitic on the other having rights, both assumptions are questionable.

Talk of moral status might be more helpful for the no-person strategy. Moral
status might come in degrees concurrent with one's developmental maturity. One
could then infer that unborn human beings have a lesser moral status than so-called
full-fledged persons. It would still not follow, however, that it is permissible to kill
them. The first claim about degrees of value is dubious since it is often based on
whether or to what extent the entity can exercise functions that the author deems are
valuable. What is questionable here is not whether such functions are valuable; they
are. The question is whether *the thing* that has those functions or will develop those
functions is an entity for which it would be wrong to kill. Focusing attention on
valuable functioning risks missing the point.

The inference to permissible killing (even if the first claim is true) is also dubious. A thing's lesser status might still be enough to justify not killing the thing.[3] A
comatose ICU patient might have lesser moral status than a so-called full-fledged
person—due to the loss of valuable functioning, but it would still be wrong to kill
that patient. If the reader is not convinced, feel free to use a different example with
a different set of functions that are lost. Consider severely autistic adults who appear
not to form a self-concept. I take it that killing an autistic person is still wrong.
Consider minimally conscious patients or patients with Cotards' syndrome

---

[3] In saying this, I am assuming for the moment that a thing's moral status can wax or wane due to
the loss or gaining of valuable functions. It is apparent given the chapters on euthanasia, that I
consider such views unjustified.

according to which they suffer delusions to the effect that they think they are dead (Ruminjo & Mekinulov, 2008). Such patients have severely depressed cerebral metabolism, particularly in the posterior cingulate and percuneus areas. Such areas are "responsible for "core consciousness" (Bauernfeind et al., 2011) and our abiding sense of self" (Charland-Verville, et al., 2013, 1998). Depressed activity in these areas compromises one's sense of self—in point of fact, Cotard's syndrome is insensitively referred to in pop culture as walking corpse syndrome. Disorders in one's sense of self would not justify killing such patients. If the exercise of certain cognitive or rational functioning is what determines one's moral status and such functioning comes in degrees, moral status would come in degrees. But individuals who wax or wane vis-à-vis one of these functions does not justify permissibly killing them.

A third concept has recently surfaced and that is the concept of violence. Abortion may be an act of violence. Kaake explains the relevance as follows, "it's clear that someone can do an act of violence against a non-person. If someone kicks a squirrel or hits a bird with a rock, we don't have to ask what kind of capacities the animal had in order to immediately understand that it was subjected to violence" (Kaake, 2013). One can perform an act of violence on something even though the something does not have rights or moral status. What matters for an act of violence is that the agent intends to frustrate the growth, maturation, and development of an organism that would *otherwise* grow, develop and mature.

Kaake's concern here is not obviated by counter-examples. One might think that it is morally uninteresting to mow one's lawn and calling that violence is an abuse of language. What is clearly not uninteresting, however, at least not without begging the very issue at stake, are actions that do violence to a developing human being. This is not to voice pro-life sympathies at this point in the narrative. It is giving voice to what abortion doctors themselves observe. Lisa Harris observes that "the violence and, frankly, the gruesomeness of abortion is owned only by those who would like to see abortion (at any time in pregnancy) disappear, … The pro-choice movement has not owned or owned up to the reality of the fetus, or the reality of fetal parts" (Harris, 2008, 77). The broader moral vision Harris subscribes to struggles to make coherent those moral ideals with abortion.

> It is worth considering for a moment the relationship of feminism to violence. In general feminism is a peaceful movement. It does not condone violent problem-solving, and opposes war and capital punishment. But abortion is a version of violence. What do we do with that contradiction? (Harris, 2008, 77).

What makes the no-person strategy so important given these initial reflections? Consider the claims about degrees of moral status. Though individuals may be deficient in one function, they are not without the most important one: namely, consciousness. More importantly, the focus on moral status misses the point of the no-person strategy because the conclusion of that strategy is that there is no person who is killed, harmed, or wronged. It's moral status, the thing killed by an abortion, would not be the status *of a person*. Considerations about moral status simply do not matter when the thing we are discussing is not a person at all.

Notice also that for Wreen's, Mclachlan's and other examples on rights, they all count as persons on some theory or other—though some cognitive gymnastics are required to incorporate the comatose into the moral community. And Kaake and Harris' commentaries on violence do not themselves entail moral wrongdoing—it matters what kind of thing suffers the violence. The no-person strategy functions to stave off each of these challenges: abortion is not impermissible violence because *a person* is not dismembered, it is not shirking a duty to care (for the developing human being) because there is no one there who is owed care, it is not a violation of an *individual's* moral status because there is no one there *who* exists, there is no person who has moral status qua person.

Regarding Wreen's example of duties towards non-persons, if taken to the extreme the idea would prohibit felling trees or mowing one's lawn. One can agree that we have a duty to the environment, but those duties are contextual with numerous exculpating conditions (e.g., it is not a duty to commute to work by bicycle even though it is good for the environment). If the no person strategy is correct, at best, pre-conscious human beings have the same moral status as human gametes, if at all. But that status would render it permissible to destroy them/it. One way to state the point is that if the no-person strategy is successful, abortion is no different than using spermicide for contraception. If something is not a person, it is not owed behaving in a way apposite for an entity that is owed moral concern.

So understood, Sect. 4.3 explains the functional account of persons which serves as a premise in the no-person argument. Section 4.4 begins the dialectic by examining the problem of underdetermination, according to which the functional view of persons is underdetermined by the cases used to motivate the view, namely, brain transplant cases. Section 4.5 entertains putative improvements on how to argue for a psychological account of the person, namely, persons are selves. Section 4.6 collates the lessons learned in previous sections vis-s-vis the de jure defects in the function argument for (pk).

4.3 The Function View

McMahan presents his readers with a quad-lemma to the question of what we are essentially. We are either souls (understanding 'soul' in an Aristotelian sense or a Cartesian one), human organisms or, equivalently, animals; or psychological entities, namely, *embodied* minds. His objections to accounts appealing to souls are separate from the dialectical progression beginning with the alternative that we are human animals.

It might be obvious that we are human animals (Olson, 2007). When I look in the mirror, I see an animal, something with a head, a body, etc. I walk, run, stretch, eat, hug, bump into others, and any number of things that only animals can do. But these are rather jejune observations, and they don't give one reason for thinking that you and I are *identical* to a human animal; I might be constituted by or housed in an animal.

Ideas that motivate the transition from animalist accounts to a functional view are the brain transplant (BT) and dicephalic twin thought experiments. These thought experiments aim to highlight the intuition that we are not identical to our bodies but rather to our psychological capacities rooted in a functional brain. Where our functional brains go, there we go also. Here is McMahan's rendition of a BT example:

> One's entire brain is extracted and transplanted into the body of one's identical twin, who has just suffered brain death and whose brain has been removed. One's brain is appropriately connected to the nerves in one's twin's body, so that after the operation a person is revived in one's twin's body who is fully psychologically continuous with oneself as one was before the operation. Most people believe that one would survive this operation and would continue to exist in what was formerly the body of one's identical twin (McMahan, 2002, 20).

Let us say that the brain of Jack is transplanted into a decerebrate body. Let us call the entity with Jack's brain Jackson. The intuition is that Jack is Jackson because where Jack's thoughts, beliefs, memories, desires and feelings go, there he goes also. At this point, McMahan has established at the very least that being a person must have something to do with being able to think, believe, remember etc.

Consider the mirror image of a BT-experiment which I will call a body transplant experiment. Imagine that Jack is a most unfortunate patient in that he suffers from lung failure, heart failure, renal failure, liver failure, several hematologic disorders, osteoporosis, and lastly, suppose that on admission he has suffered severe burns over 90% of his body including his face. He basically needs new skin, bone, and bone marrow; whole blood transfusion, a kidney, a lung, a heart, and so on. Happily, the hospital has matches for each of his organ systems. They transplant his entire set of organs, his brain is the only thing that is not transplanted. The intuition is that Jack is the same person before the transplants as after. The only thing that did not change is his brain and the functioning that it supports.[4]

Once a largely psychological account of the person is motivated, McMahan motivates an embodied mind account. For McMahan, the specific contents of one's psychology—her beliefs and memories—are sufficient but not necessary for being a person. Considering cases of deprogramming whereby a person loses his/her specific memories and beliefs such as with progressive dementia are still cases in which we say the person exists. Of course, we might say about a loved one who is demented that he is 'no longer himself,' or that he is 'gone'. These adjectives refer to the loss

---

[4] One can see an important assumption according to which, what is understood by "animal" must be, according to proponents of the psychological view, that it is everything about a human organism *except* the brain. This is important because the alternative hylomorphic view of the person understands that persons must have rational potencies, but that the person is the entity or thing that has these potencies. Thus, the person can pre-exist the actual exercise of these potencies. In this regard, brain transplant examples are counter-examples to those views of the person that exclusively focus on the organism understood as everything about an organism *sans* its functional brain. I do not think that animalists believe this, but maybe they do. What I am more certain of is that the hylomorphic view is not committed to a narrow conception of the body or animal (Eberl, 2020; Hershenov, 2011; Broackes, 2006; Hacker, 2009; Gorman, 2024).

of the person's specific narrative. A loss or change in one's specific narrative is not necessary for being a person as can be appreciated by considering cases in which someone is struck by lightning and their personality changes, or by considering cases of transworld identity. Consider someone who is about to make a major life choice and makes a different one than what was actually chosen, resulting in a radically different life narrative. (Suppose, for example, that this major life choice concerns accepting or rejecting a moral or religious viewpoint so one's core beliefs and desires would be different). Intuitively, we would still say that it is you who converts. So, biography is not determinative of one's ontological identity qua person.

The embodied mind account wishes to balance our intuitions on the BT experiments with the cases of deprogramming and narrative change. The account

> stresses the survival of one's basic psychological capacities, in particular the capacity for consciousness. It does not require continuity of any particular contents of one's mental life. This allows that one may survive the deprogramming of one's brain and that one continues to exist throughout the progress of Alzheimer's disease, until the disease destroys one's capacity for consciousness (McMahan, 2002, 68).

Summarizing the embodied mind account, McMahan notes that,

> There need be only enough physical and functional continuity [of the brain] to preserve certain basic psychological capacities, particularly the capacity for consciousness. This, I believe, is a sufficient basis for egoistic concern; it should, therefore, be a sufficient basis for identity, other things being equal (McMahan, 2002, 68).

Functional continuity is defined as "the retention of the brain's basic psychological *capacities*" (2002, 68, emphasis in the original) and physical continuity requires "either the continued existence of the same constituent matter or the gradual, incremental replacement of the constituent matter over time" (McMahan, 2002, 68).

Care should be exercised in interpreting McMahan on these points. McMahan considers brain-in-the-VAT scenarios (BIV) (for which McMahan—and I—lose the intuition that the person is preserved) and also dicephalic twin cases (for which McMahan—and I—have the intuition that there are two different persons, but they share the same body). One might get the impression of contradictory intuitions; if the BIV is not a person even though it is a functional brain, but the dicephalic twins are two persons because they have two functional brains, then our intuitions are unreliable regarding the importance of a functional brain to that of being a person.

We can bring clarity by focusing attention on the BT intuition, namely, Jack is Jackson. What did the brain transplant do to Jack? One plausible interpretation is that a person and that person's animal existed at time t_1. After the transplant occurs, that same person exists but with a different animal. So, $(P_i \, \& \, A_a)$ exists at t_1. After the transplant, $(P_i \, \& \, A_b)$ exists at t_2, and the following identity relations hold: $P_i = P_i$; and $A_a \neq A_b$. Since personal identity is preserved but animal identity is not, $[P_i \neq (A_a \,^{\vee} A_b)]$.

On this reading, the transplant examples are still compatible with either of the following claims:

- Every person is an animal (maybe not *this* animal, but *an* animal).
- No person can exist *apart from* an animal.

Think of musical notes and pitch. Every note has a pitch, but note and pitch are different things. Likewise, agreeing that Jack is Jackson is compatible with believing that all persons must be embodied. The ontological striptease (Dancy, 1978, 395ff.) we are invited to perform with brain transplant thought experiments should not suggest that we can exist without an organism, even if we are not this organism. McMahan is sensitive to these points as reflected in both the brain-in-vat scenarios and the dicephalic twin examples. Even though dicephalic twins share a body they are, on the one hand, different persons due to having different centers of consciousness and functional psychological capacities; but on the other hand, persons must be embodied, even if that body is shared. In essence, BT experiments tell us that there is one mind that exists in two bodies, with the intuition that there is one person. Dicephalic twin cases tell us that there are two minds in one body, with the intuition that there are two persons. The thought experiments mutually support each other.

I take the embodied-mind account of the person to be the most plausible species of a psychological theory.[5] The chief motivation for this view is rooted in the brain transplant-dicephalic twin thought experiments on one hand, and the deprogramming-Alzheimer's cases on the other. What the BT-experiments intend to show is that our psychological properties are fundamental to our identity. The procedure is to use egoistic concern as a heuristic to identify what it is that you and I are essentially. The procedure asks what changes could occur to me for which I would remain concerned about my survival? The deprogramming and Alzheimer's cases motivate the idea that it is not the specific psychological *contents* that ground identity (or egoistic concern) but a preservation of one's basic psychological capacities. In a way, these two sets of thought experiments work together by limiting the scope of each. Whereas the BT experiments motivate the view that specific contents of one's psychology (e.g., specific beliefs and memories) are sufficient for being *this* person,[6] deprogramming thought experiments motivate the view that specific contents are not necessary for being *a* person. The two experiments taken together suggest that

---

[5] Lynne Baker's constitutionalism deserves mention as being on par with McMahan's account. See Lynne Rudder Baker (2000). The differences between embodied mind and constitutional accounts will not affect the argument in this book.

[6] There is a complication regarding teletransportation thought experiments according to which one's contents are duplicated (somehow) and inserted (somehow) into a different body/brain. The intuition we are supposed to have is that the duplication of content duplicates the person, i.e., there are two persons with the same content. For those who have such an intuition having a specific set of beliefs and memories is *not* sufficient for picking out *this* person. McMahan and others do not find such thought experiments persuasive; and neither do I for the reasons McMahan and van Inwagen note. See McMahan, 2002, 56–59. In addition to McMahan's observations, I am inclined to think that something like belief transfer is either irrelevant or impossible. It is true that you and I can have the same belief in terms of its propositional content. Both of us can believe that 'it is raining outside'. But it is unintelligible how *you* can have *my* belief that it is raining outside. If what gets duplicated is simply the content, we did not need a thought experiment to tell us that. If what gets duplicated is *my* belief, we are asked to imagine the incoherent.

what is necessary and sufficient is the preservation of one's capacity for consciousness rooted in a functional brain.

How does such a view of the person justify the permissibility of abortion? The question is a species of a more general question concerning what counts as wronging *someone* (Chappell, 2011). To answer the general question, we need to know what persons are and when they come into existence. The embodied mind account answers that persons are embodied minds, and they do not come into existence until the human brain can exercise psychological functioning. Such accounts have focused on justifying the permissibility of early-term abortions that McMahan defines as "…an abortion that is performed prior to…the point at which the fetal brain acquires the capacity to support consciousness…." (2002, 267). For the sake of argument, assume that this developmental milestone occurs at around 22 weeks gestation. Prior to 22 weeks gestation, the embodied mind account delivers the judgment that there is no person there to be killed. Constitution views deliver the same judgment. Baker comments that "[w]hen a human organism develops to the point that it can support a first-person perspective, a new entity—a person— comes into existence" (Baker, 2000, 155). Where there is no person, there can be no rights (proper to persons) that might be violated by killing it. Actual psychological functioning is necessary to count as a person,[7] and being a person is necessary to be a member of the moral community.[8] If an abortion were to occur prior to the exercise of psychological functioning, that abortion would not harm anyone, for there is no one there to harm. If there is no harm, there can be no wrongdoing. Therefore, if abortion occurs prior to the exercise of psychological functioning, abortion is permissible.[9]

4.4 The Problem of Underdetermination

The concept of underdetermination is notoriously labile and imprecise (Laudan, 1990; Seidel, 2014). The basic insight, however, is plausible. It is the claim that "the evidence available to us at a given time, may be insufficient to determine what beliefs we should hold in response to it [the evidence]" (Stanford, 2017). For my

---

[7] Baker comments, "When a human organism develops to the point that it can support a first-person perspective, a new entity-a person-comes into existence" (2005, 36). And a first person perspective is the ability "to think of oneself without the use of any name,…; it is the ability to conceive of oneself as oneself, from the inside, as it were" (Baker, 2005, 28).

[8] To be precise in this chapter, inclusion in the moral community refers specifically to the right not to be killed.

[9] David DeGrazia offers the following appraisal of McMahan's argument, "Surely this is one of the most elegant defenses of abortion (or at least early abortions) ever conceived." (*Human Identity and Bioethics*, 280). Dean Stretton observes that McMahan's *Ethics of Killing* is "the best work ever on abortion and related topics" (Stretton, "Critical Notice: *Defending Life: A Moral and Legal Case against Abortion Choice*," *Journal of Medical Ethics* 34(2008), 793–797 at 797).

purposes here, it is sufficient to distinguish four versions of the underdetermination thesis:

1. Theory T is underdetermined by a set of evidence $\{E\}$, *iff* T and T_i are both *compatible* with $\{E\}$ and $T \neq T_i$.
2. Theory T is underdetermined by a set of evidence $\{E\}$, *iff* T and T_i both *explain* $\{E\}$ and $T \neq T_i$.
3. Theory T is underdetermined by a set of evidence $\{E\}$, *iff* T and T_i are both *justified* by or *supported* by $\{E\}$ and $T \neq T_i$.
4. Theory T is underdetermined by a set of evidence $\{E\}$, *iff*, T is *not* supported by or justified by $\{E\}$.

The first three versions concern underdetermination of theory *choice* because in an important sense, they are not each underdetermined, since they each are compatible with, explain, or are justified by the same evidence set. But which one should be believed is underdetermined. The fourth concerns underdetermination of a theory. Which version is relevant to my argument in this section depends on how one interprets the role of thought experiments. I argue here that the function view is underdetermined in the sense of either 1, 2, 3, or 4. And if it is underdetermined by at least one, it fails to justify (pk). To be clear: Theory T is the function view, theory T_i is something like a hylomorphic account of the person, and the principal evidence is our intuitions on thought experiments.

The reason for this precision about the notion of underdetermination is because it frames a more ecumenical rejoinder to the function view's case. Theorists who are sympathetic with an animalist or hylomorphic view[10] of the person seem tempted to respond to brain transplant examples by denying the intuition that the person survives the transplant. Snowden, for example, says that "I shall suggest one possible initial reply, according to which we are under no obligation to believe the fourth premise" (Snowdon, 2014, 203). For Snowden, the fourth premise of the brain transplant argument is stated as, "the resulting subject sustained by (or realized by) this system [the brain plus the new organism] would be *the same subject* (or person) which that brain previously sustained" (Snowdon, 2014, 202, emphasis added). We should not agree, Snowden claims, that brain transplants are person-preserving. Likewise, Wiggins (2001, 211) says the following,

> I claim that the creatures or artefacts that these further developments offer to our speculations [brain transplant and brain fission cases] will need to be seen as analogous to clones, copies, reproductions, models—as all of a piece with things that approximate to their originals, or simply behave as if they were their originals.

On Wiggins' view, Jackson is *like* Jack, but is not identical to him. Jackson is a reproduction or copy of the original. Personal *identity* is not preserved. Finally,

---

[10] Such views are considered in the next chapter. Here it is enough to understand that such views understand us to be substances or entities with a rational nature. Having immediately *exercisable* rational capacities is not necessary to be a person.

Hacker observes that "the science fiction [of BT experiments] does not show that if A's brain is transplanted into B's decerebrate skull, then the resultant person, let us call him Aby, is actually A" (Hacker, 2009, 306).

These views deny the intuition that many of us have, namely, where Jack's functional brain goes, *he* goes. Where one's thoughts, desires, beliefs, and more importantly, center of consciousness go, there *that* person goes. These authors appear to deny that brain transplants can be person-preserving.

The position I defend in this section is that one need not deny the central intuition of the brain transplant example to take issue with its justification for (pk). (This is not to suggest that I disagree with the authors just quoted, but I need not defend those views here.[11]) Rather, I aim to defend each of two claims:

(A)   The intuition that brain transplant examples invite us to have can be explained, justified, or accommodated by non-function views of the person. These views understand persons as coming into existence at or immediately after conception. This claim concerns underdetermination in the sense of 1, 2, or 3.
(B)   Brain transplant examples do not provide any intuitive evidence about *when* persons come into existence. This claim concerns underdetermination in the sense of 4.

If either (A) or (B) are true, the justification for (pk) is undercut.

(A) and (B) are logically related roughly as follows. If a theory can accommodate the transplant intuition *and* it tells us that persons come into existence prior to exercising their psychological functions, BT experiments do not tell us when persons come into existence. If the transplant intuition is logically consistent with different views about when persons exist, then it does not provide evidence about when persons exist. Conversely, if a view does not tell us when persons come into existence then that view would be logically compatible with the transplant intuition, insofar as BT experiments presuppose that Jack is an adult person at t_1 and we are asked if he is still that same person at t_2—nothing is presupposed about when Jack *first* existed. Though they are logically related, they are not making the same claim. So understood, I consider them separately.

The following explains roughly what is at stake in defending (A) and (B). For (A), if a piece of evidence is compatible with theory$_1$ and theory$_2$, it does not exclusively *support* theory$_1$. If brain transplant thought experiments invite us to have intuitions that can be accommodated by theory$_2$ and theory$_2$ has us coming to

---

[11] For instance, I think that Wiggins' analysis is correct and obviously so when we consider fission cases. Suppose that Jack's brain is divided and distributed to two decerebrate but functional bodies (say, Jackson$_1$ and Jackson$_2$). It would deny the logic of identity to say that Jack is Jackson$_1$ (and for the same reasons, is also Jackson$_2$), since Jackson$_1$ $\neq$ Jackson$_2$. It seems perfectly sensible, however, to say that Jackson$_1$ (or Jackson$_2$) is a copy or reproduction of Jack, but not identical to Jack. Transplants, then, are not identity preserving. One might carve out an exception in fission cases, but that strategy appears ad hoc. Likewise, Hacker's commentary on brain transplant examples is quite sensible once we are clear about what mind, body, brain, and person mean.

existence at or close to conception, the evidence of such thought experiments under-determines the defense of abortion.

For (B), I argue that the brain transplant thought experiments do not provide intuitive evidence that tells us when persons come into existence and we need to know that to justify abortion.

The function view is a premise in the no-person argument for (k). The de jure defect of that argument is that the reasons for the function view fail to motivate, justify, or support the idea that you and I are not persons at or shortly after conception. The no-person view does not justify (pk). Given the evidence of BT-experiments, there may (or may not) be a person there who is wronged by being killed. (pk) does not follow from nor is it supported by the reasons for the function view.

4.4.1 Underdetermination: Compatibility with Other Views and a Defense of (A)

David Hershenov supposes that a hylomorphic view of the person does not require rejecting the transplant intuition, namely, that Jack is Jackson, but it can consistently reject the idea that you and I exist only when we can immediately exercise psychological functioning.

He notes that for Aquinas, to consider one hylomorphic account, the human animal "is a distinctive animal due to its capacities of intellect and will" (Hershenov, 2011, 469). Because a functional cerebrum is what confers on a human being the powers of intellect and will, one would be correct to think that where the cerebrum goes, the person goes. "If the soul provides the capacity for rational thought, and the person will be found where his or her soul is, then one has some reason to claim that the soul and the person have moved when the cerebrum does" (Hershenov, 2011, 469). A human animal, according to Aquinas, is not a mere organism, but a kind of organism, one that has characteristic powers, morphology and developmental plan. We tend to think of organisms as things with hearts, liver, lungs etc.. But for those in the hylomorphic tradition, paradigm organisms are trees, dogs, and human beings. These are different kinds of organisms, and the differences follow differences in powers, morphology and developmental trajectory. Important for understanding the root difference between a hylomorphic view or animalist view on the one hand and a functional view on the other is how the former understand substance. A human substance is an individual human being. This being is primary in the sense that walking, thinking, or having a color and shape, and other properties require the substance or entity to exist. Thinking requires a thinker. To foreshadow a bit, consciousnesses (if I may use the plural) do not have independent existence. Being conscious is a state or capacity of thing.

In that regard, the key difference between a hylomorphic view of the person and the embodied-mind account concerns the metaphysical importance of *having* versus *exercising* a rational capacity. For the embodied-mind account to function as an argument for the permissibility of abortion, being a person requires an exercisable capacity for rational activity. For the hylomorphic account, being the kind of thing that by its own self-directed development manifests such capacitiesis by that fact a person. The reason this entity has such capacities in the first place is that it has a rational nature. The nature of a substance S explains why S is the way it is, develops the way it does, and has the capacities it does. On such a view, I can have rational powers, qua human being, before being able to exercise them. An eaglet has the ability to fly because it is a bird, even if it has not developed wings yet. A human being has rational abilities because she is a person, even if those powers are not actually functioning yet. So long as one holds to the common-sense intuition that you and I are things—entities that have powers but are not identical to a specific power or capacity—you and I can exist prior to the functioning of a specific power.

Theories of the person that emphasize *having* versus *exercising* rational potencies is an emphasis that is small enough to accommodate the intuition that Jack is Jackson per the original brain transplant experiment; but big enough to accommodate a view of the person according to which persons come into existence at or shortly after conception.

The inductive case for claim (A) builds if we add to the original BT experiments additional permutations. These permutations aim to supply intuitive evidence that personal identity can *also* track continuous biological development of the individual.

Consider permutations of Snowdon's brain-budding cases (2014, 224–225). Suppose we have brains that deteriorate precipitously but leave behind a cluster of pluripotent stem cells that reproduce the cerebrum every night.[12] The reproduced brain is identical in functioning and content to the deteriorated brain. This is not too fanciful, as our cells replace after a few years, and even sleep involves an elaborate neurological process of reorganization and learning (Stickgold, 2005).[13] Just imagine accelerating the process of cell replacement after apoptosis, broadening the processes of reorganization/regeneration, and shortening the time window in which these processes are completed. Readers are invited to shorten this window, if they like, until they have the intuition that such a process is person-preserving. Suppose, for example, that the process of cerebral collapse into a cluster of pluripotent stem cells followed by regeneration takes 1 hour. Suppose that this cycle of cerebral deterioration and apoptosis followed by regeneration is as biologically determined as metabolism, growth, and maturation is in the actual world. Intuitively, it is Jack who wakes up in the morning; and this for the same reasons we say that Jackson is Jack,

---

[12] Suppose that all the stem cells are required for such regeneration, and that splitting them up to does not yield two people. Such a qualifier is not fanciful either. Consider splitting the cells of the human embryo at the blastocyst stage. Doing so does not create twins, it destroys the human embryo.

[13] In reality, mental content does not remain the same before and after sleep since sleep is essential for learning and memory reorganization. But for the purposes of the example, assume the content is constant.

namely he has the same beliefs, desires, memories, etc. Now consider the following cases:

Regeneration + transplant: Suppose that after Jack falls asleep and his cerebrum returns to a cluster of pluripotent stem cells we transplant such stem cells into a decerebrate body. Does the person go with the stem cells? If we think that it is Jack who wakes up in the morning (in the formerly decerebrate body), it does not seem to matter which animal or organism he wakes up in. If this is the way cerebrums maintain functional integrity through apoptosis and reorganization, then transplanting Jack's pluripotent stem cells into another body transfers Jack to that body.

Regeneration + destruction: Now suppose that after Jack's former brain deteriorates and before the pluripotent stem cells regenerate a fully functional cerebrum, one removes such stem cells scatters them and, for example, isolates them for stem cell research. Doing so effectively ends Jack's existence.

In the case of Regeneration + transplant, Jack goes where the pluripotent stem cells go for the same reasons that Jack goes where his brain goes in the original brain transplant experiments. Since the reasons are symmetrical, we are committed to saying that it is Jack who wakes up in the morning in the Regeneration + transplant permutation. This suggests that our intuitions regarding personal identity track developmental processes as well. The original BT-experiments consider that an actual functional brain is transplanted into a functioning but decerebrate body. Personal identity is preserved. In these regeneration/developmental permutations, a brain in its primitive stem cell state is transplanted into a functioning but decerebrate body. Personal identity is also preserved.

In the case of Regeneration + destruction, we are inclined to think that Jack's existence is ended. If one is not sold on this idea, consider that the one who scatters the pluripotent stem cells is Jack's arch enemy. Would the arch enemy think that his actions would be *successful* in ending Jack's existence? Certainly she would. Furthermore, if you think that Jack survives in the transplant permutation but he does not in the destruction permutation, you should think that the arch enemy's action of scattering Jack's stem cells *killed* Jack since that act of scattering is the difference maker, i.e., it is what killed Jack.

The point in introducing these examples is to suggest that our intuitions regarding personal identity track *natural biological developmental processes*. They illustrate that certain processes of biological development are compatible with the identity of persons. When we import features of biological development into a transplant case, it remains intuitively plausible that Jack is preserved—in Regeneration + transplant permutation; and that Jack is killed—in Regeneration + destruction permutation.[14]

---

[14] Baker (2005) may demur. She thinks that we come into being when we demonstrate a rudimentary first person perspective and this requires having a "developmental preliminary" to a robust first person perspective. The context is important: she is trying to distinguish between being a human infant and being a chimpanzee even though both have rudimentary first-person states. Baker is

A final permutation suggests that Jack exists when the elementary conditions for his psychological capacities first surface.

Generation: Suppose this is Jacks first developmental point in which his incipient pluripotent stem cells cluster in the base of Jack's spinal column for the first time. The process is as above, such stem cells generate a functional cerebrum. Assume, as you like, that Jack's morphological features are fully formed: Jack has eyes, ears, a head, cardio-pulmonary functioning, limbs etc. and they are all developed immediately after conception. Of course, Jack does not yet see or hear anything, but he has a heartbeat.

Intuitively, if we say that Jack is destroyed in Regeneration + destruction at the point that Jack's brain is in a pluripotent state, Jack exists in the Generation case too. If Jack exists when his cerebrum is in a pluripotent stem cell state (as with the case Regeneration + transplant), Jack exists in Generation. The reason: at the time that Jack is in a pluripotent state, he is identical in terms of kind and powers across the Regeneration and Generation cases. What more is *necessary* to count as a being Jack?

It is a non-essential feature that Jack *did* exercise psychological powers in the Regeneration permutations. The reason prior existence cannot be a necessary condition for being Jack is that it is indiscriminate in other cases. A dead patient did possess consciousness, but the patient does not exist. A temporarily comatose patient did possess consciousness but clearly does exist. Past possession of psychological functions is not a feature that itself differentiates between present ontological statuses. To be clear, I am not suggesting that the exercise of psychological powers is a superficial feature for reidentifying Jack. Rather it is the most important feature. But their *occurrent exercise* is not necessary for *being* Jack—otherwise we have to say that Jack does not exist in the Regeneration cases when his brain is in a pluripotent state.

For reasons that are opaque to me, the claim that past existence is required is popular. Tooley, for example, says that temporarily comatose patients are persons because they did possess consciousness, but the fetus did not. Likewise, he would say that Jack exists in Regeneration, but not Generation. Tooley claims that an organism "cannot have a serious right to life [i.e. cannot be a "person"] unless it either *now* possesses, or *did* possess at some time in the past, the concept of a self as a continuing subject of experiences and other mental states together with the belief that it is itself such an entity" (1972, 82, emphasis added). Mary Anne Warren

---

right to ground the distinction in their different developmental trajectories, i.e., "developmental preliminary." But if developmental potency matters, why not say that the human embryo counts too, even though he has a remote capacity for first-person states? Baker replies: "I do not believe that remote capacities suffice for making anything the kind of thing that it is" (2005, 35). This comment, however, reveals a misunderstanding. The capacities do not explain or cause the thing to be the kind of thing it is. The opposite is the case. The nature of a thing explains why it has the capacities it does. A thing has the capacities it does, including developmentally indexed ones, *because of* the kind of thing it is. (On these latter points, see Heaney (1992), Henry (2005), Condic (2011), Gomez-Lobo (2002), Eberl & Brown (2011), Oderberg (2008), Scaltsas (1994), DiSilvestro (2010). See Eberl (2020) for a detailed analysis of Baker's Constitution view.

also makes a similar claim in which an actual or past exercise of consciousness is necessary to count as a person (1992).

Schwarz and Tacelli argue that these accommodations are underwhelming. "Is it any less plausible to say: "An organism can be called a person if it either *now* possesses or did possess at some time in the *past* or will possess in the *future*, the concept of a self?" (1989, 89). Tooley's response, they note, is rather ad hoc. "But why should that count for so much? Why should a great moral chasm yawn between "did, but does not" and "does not, but will"? The precision seems ludicrously *ad hoc*" (1989, 90). One would need an explanation for why there exists an ontological and moral chasm between "did but does not," and "does not but will" vis-à-vis psychological functioning.

Here is another reason for thinking that this response is underwhelming. Consider Jack in one of the regeneration cases in which his brain has returned to its pluripotent state. Index this point as time T1. Jack exists at T1. Now consider Jack in the Generation case according to which his brain is in an identical pluripotent state at T1. In the Generation case, either Jack exists at T1 or he does not. If we say he does not, and yet he is identical to Jack at T1 in the regeneration case, then what else is he missing? It seems we have to say that he's missing existence at time T0 (a prior time). But this position would entail that past existence is required to exist. In order to exist, one must exist previously. One can easily see a regress forming if that is correct. There would be no first existence; only eternal things would exist. But that cannot be right.

Maybe the criterion for coming into being is having one's first episode of consciousness. But if a state of consciousness is required to be, Jack does not exist at T1 in the Regeneration case either. If a capacity for consciousness is required, Jack has that in both Regeneration and Generation at T1—because of the kind of thing he is. To be consistent, we have to say that Jack exists at T1 in Generation as well.

We can get at the same conclusion by contrasting Generation with Destruction. The reason we think that Jack was killed in the Destruction permutation is that Jack *would* develop his cognitive apparatus complete with his thoughts, beliefs and intentions. The very idea of killing or ending one's life is parasitic on the idea that *that life* would continue had the act of killing not taken place. If we say that Jack was killed in Destruction, and he is identical to the-thing-at-T1 in Generation, Jack is that thing-at-T1 in Generation; Jack exists in Generation. The point is to reveal that our intuitions on killing do not consider past existence. What matters is that this life ends at T1. The fact that Jack will possess the full array of his cognitive apparatus is required for the concept of killing to apply (though it might not be sufficient). The fact that Jack's survival morally matters suggests that Jack exists in the Generation case.

To summarize this sub-section, brain transplant thought experiments do not tell us what a person is, viz., what is the nature of being a person. At best, they advert our attention to the evidence used for re-identifying a person at one time with that same person at another time. It is assumed that there is a person at both times. This observation is not at all unusual. Eric Olson observes that "[t]he traditional problem of personal identity is not what we are, but what it takes for us to persist. It asks what is necessary, and what is sufficient, for a person existing at one time to be identical with something present at another time: what sorts of adventures we could survive…" (2003, 323).

The Regeneration and Generation thought experiments are meant to modify the original brain transplant thought experiments with a feature of biological development and self-directed determinacy. Once we add such features, our intuitions follow, so to speak, developmental trajectories as well. That is, *we are inclined to think that Jack's identity is preserved through biologically determined development typical of his kind*. It is true that the modified thought experiments still make Jack's psychological functioning an essential feature of Jack, but the permutations show that the *actual* functioning is not necessary for being a person. As Schwarz and Tacelli note (1989), one can *be* a person without *functioning as* one. Brain transplant thought experiments do not give us a reason for thinking that the pre-born human being is not a person. The intuitions that the original BT experiments invited us to have can be accommodated by a hylomorphic account of the person, and the regeneration and generation permutations above provide direct intuitive evidence that personal identity can track biological development. At the very least, we have a case in which two different theories of the person are either compatible with, supported by or equally explain the same set of evidence (viz., our intuitions on cases of personal identity). This result is enough to justify that the function view suffers underdetermination in either of sense 1, 2, or 3.

4.4.2 Underdetermination: When Do Persons Exist

That Jack survives the transplant *after* he has developed psychological powers does not entail that he does not exist *prior* to the development of those powers. Consider a brain transplant case occurring at the embryonic stage of development with twins. Suppose that embryo E_a exudes normal cerebral development but embryo E_b is anencephalic; and that E_a's cerebral development has not reached a point at which it can support consciousness yet.[15] E_a however, suffers from severe dysfunctional renal development, but E_b has a functional body in all respects. E_a's tiny cerebrum is transplanted into E_b's decerebrate body, a procedure that is certainly metaphysically possible. Is E_a's life saved by this procedure? Does E_a survive? We might ask more generally of the maternal-fetal surgeon doing the procedure, what clinical reason exists for doing it? The most sensible answer is that the surgeon is trying to save E_a's life.

---

[15] In normal human development, the two spheres of the cerebrum can be discerned at 3 weeks gestation, with brain waves and reflexive activity discernible at 5 weeks. If we suppose that sentence is a marker for some primitive level of consciousness, it is still the case that there is a two-week window in which this thought experiment might take place.

Controversy might exist if one infers from E_a's life is saved, to a *person's* life is saved.[16] But to motivate that inference we may fast-forward the case wherein the human being at such an embryonic stage continues to develop, is born, and grows into childhood. The parents inform him, call him Anthony, that he would have died had they not transplanted his primitive cerebrum into his twin's decerebrate, embryoid body. Intuitively, what the parents say is plausible and my interlocutors should think so too—these are our intuitions on prenatal harm cases discussed in Chap. 3. DeGrazia (2005, 289ff.), McMahan (2002, 211ff.) and others accept that if the child survives, the child is harmed by, for example, progressive neuron removal, or fetal alcohol syndrome, or the mother's smoking habit while being pregnant. If we say in the prenatal harm cases that the adults can truthfully say 'I was harmed by the [enter prenatal intervention],' Anthony can say that he was saved by the surgery. Just as for prenatal harm cases, the intuition is that the person is harmed when the prenatal harm occurs; so too in the embryoid transplant case according to which the person is saved when the intervention occurs. It is Anthony who is spared the harm. So, everyone should agree that Anthony's life was saved by the cerebral transplant.

The point of the example is to challenge the idea that an *occurrent*, functional brain is required for being a person. By the term 'functional' proponents of the function view mean occurrent or the immediate ability for occurrent consciousness. What the thought experiments here suggest is that persons can exist prior to such occurrent states. So long as one has a brain in a primitive developmental state, as does Anthony, he exists. Importantly, this conclusion is compatible with what we learned from the BT experiments; we can say that Anthony is saved by the surgery *and* that Jack is Jackson in the original cases. The intuition on the embryonic brain transplant permutation and the explanation we give for that intuition are compatible with what we say on the BT experiments. I emphasize this point because readers might be tempted to obviate the lessons learned from the embryonic permutations by asserting that the embryoid human beings are not persons. That claim may or may not be true. The chief point in this section, vis-a-vis the problem of underdetermination, is that there is no reason for that claim; *no reason that can be inferred from the original BT experiments.*

If one is not convinced, one last attempt might be promising. Consider yet another permutation of an embryonic brain transplant, but in this case there is a brain swap with the additional feature of an impairment. Suppose there are fraternal twin human beings at the embryonic stage E_a and E_b (named Anthony and Beatrice respectively). Anthony and Beatrice at the embryonic stage are neuro-typical but

---

[16] The controversy is dependent upon understanding 'embryo' as a noun. The term embryo can be used as a noun to refer to a thing as in "the embryo" or "an embryo." It can also be used in an adjectival sense to describe the human being at a particular stage of development as when we say "embryonic stage" of a human being. Otherwise stated, embryo can function as a phase sortal or as a substance sortal (Wiggins, 2001). I see no reason for favoring noun usage. Does the embryo die when the fetus comes into existence? Surely not. In any case, the thought experiment is not meant to *presuppose* that embryo should be understood as a phase sortal. Rather, the intuition is simply that the surgeon saves a life of the human organism.

Beatrice suffers from congenital malformation of her limbs rendering her paraplegic in the future. Suppose, as with the previous transplant case, we swap the pluripotent stem cells that form the cerebellar neurological system. Anthony's stem cells are placed into Beatrice's former embryoid body, and vice versa. The swap is successful, gestation occurs as normal, Anthony and Beatrice are born and grow into childhood. They are informed of the swap. Intuitively, Anthony could truthfully say that "had it not been for the swap I would not be paraplegic." If this is true, Anthony goes where his *rudimentary* potencies for psychological functioning go.

Again, consider our intuitions on prenatal impairment cases. The intuition on the brain swap cases should be that Anthony was impaired *when* the swap occurred. If we say in the prenatal harm cases that the adults can truthfully say 'I was harmed by the [enter prenatal intervention],' so too we should conclude that in brain swap plus impairment case: the one who is transplanted into the paraplegic embryoid body *is* post-transplant, paraplegic. If one agrees that in prenatal impairment cases the agent is harmed by the intervention, transplanting that brain elsewhere is compatible with saying that the person exists at the time that the impairment was done.

The central idea is that the intuitions we have on prenatal harm cases, brain transplant cases and their permutations are all coherent vis-à-vis a hylomorphic view of the person; viz., one that holds that persons come into existence at or shortly after conception *and* that personal identity tracks developmental processes as well. None of the thought experiments considered so far *exclusively* motivate a view of the person necessary to defend the no-person strategy. The intuitions here are like empirical facts, and the different philosophical accounts of the person are like scientific theories. And like many stories in the history of science (Kuhn, 1996), the facts do not adjudicate or exclusively motivate one theory over another.

Before leaving this section, one last argument is worth considering. McMahan introduces an example aimed to show us that "I" does not refer to an organism, at least, not a very immature organism. Before having us consider brain transplant examples, he has us consider an often overlooked and persuasive thought experiment.

> Imagine, for example, that in some of us the process of biological development were somehow reversed. Those to whom this happened would begin to grow younger, in biological terms. Eventually they would revert to being babies and thereafter would have to be placed in artificial wombs to survive. As their brains reverted to the infantile and fetal stages of their development, their mental lives would become increasingly rudimentary and would eventually disappear altogether when their brains ceased to be capable of supporting consciousness (McMahan, 2002, 29).

McMahan asks us to consider when in this continuous regression one ceases to exist. Although the exact point at which one passes out of existence may not be determinate, I find McMahan's assessment plausible. "I find it impossible to believe that I would still be around when what we may neutrally designate as my organism had been reduced to a microscopic network of cells from which any possibility of consciousness had vanished" (McMahan, 2002, 29).

I agree with McMahan, but we find this assessment plausible for different reasons: I find it plausible because the network of microscopic cells has wholly lost the potency for rationality. It (whatever we can call it at this stage of disintegration) is along a trajectory of disintegration and nonexistence. The developing human being, however, has precisely the opposite potencies and developmental trajectory. The human being at the embryonic stage has the potency for rationality and volition, and (given a supportive environment) will exercise these potencies in the future. I happen to think that this asymmetry is ontologically important, McMahan does not. This disagreement itself is not the important point. The important point is that the case itself does not adjudicate which of our respective explanations is correct.

McMahan finds his explanation plausible because there is no actual state of consciousness that the microscopic network of cells can manifest, and I agree. There is nothing, however, in the description of the thought experiment that precludes my explanation from being correct. Here too, the hylomorphic view can accommodate the intuitions on these thought experiments. The intuitions these thought experiments invite us to have do not, therefore, exclusively support a psychological view of the person.

4.5   The Self: Consciousness *Is* the Person

Some might think that one does not need thought experiments to motivate the view that you and I are psychological entities. That persons are mental or psychological entities is a principle or primitive point of departure. Baker (1995, 491) supposes that "a person… is a being with a capacity for certain intentional states like believing, desiring, intending, including first-person intentional states." No thought experiments are offered in support of this view, it is taken as intuitively obvious. The intuitive idea is this: that thing that is conscious is me and I am a person. The ability to assert "I…" is a *necessary* marker for personhood.

4.5.1   What Exactly Is the Self?

There is, however, some confusion in stating what this view amounts to. For example, Baker says towards the end of the paragraph from which the above quotation draws, that "in order to be a person, any being must *have* first-person intentional states" (Baker, 1995, 492, emphasis added). The problem is, of course, the first quotation suggests that being a person requires only having a *capacity* for intentional states, and the latter quotation suggests that one must have occurrent intentional *states*. The problem is that capacities might be possessed but unable to operate due to disability, injury (comatose in the ICU), or temporary suppression (intoxication or sleep). Is it enough to have the capacity for consciousness or does one need to be in a state of consciousness?

This mistake might be an excusable cognitive slip. Baker is giving expression to the idea that the person is the proper subject of psychological states or attributes that are connected by an identity preserving causal relation, such as interconnected memorial states. The self is the subject of such psychological states and attributes. Though I share some sympathy with this view, we still need a reason to reify the self into a separate entity within or constituted by the human being, rather than treating it as a developed power of the human being. If consciousness is a developed power of the human being and you and I are not a power or capacity, you and I cannot be *identical* to consciousness. I might *be* conscious (of my environment, of what I'm reading, of the person with whom I'm conversing and so on); but I'm clearly not identical to episodes of consciousness. I may have them, but I cannot be them. We need a reason to think that the so-called self is an entity rather than just an abbreviated way of talking about the psychological abilities of a person.

In any case, the chief problem for Baker is her very first move in articulating the Constitution view. Whereas the *relation* of constitution is plausible enough and understandable (viz., the human animal is a person as the bronze is a statue; where the 'is' in those claims is the 'is' of constitution, not identity (Baker, 2000)). What is undermotivated is thinking that the person does not exist *until* the human animal has first person mental states. By now the reader can anticipate the next question: why reify the person into a different entity simply because the human being acquires the ability to reflect on his or her own mental states? We do not think this way for other animals with developmentally indexed potencies. The eaglet is this bird before it can fly, and it is this same bird after it can fly.[17]

To gain more clarity, we might have to be clear about what consciousness is. Following Hacker (2013) we might distinguish between intransitive and transitive consciousness. Transitive consciousness corresponds most closely to common

---

[17] One might suggest that the answer is obvious: having first person mental states is a non-natural property. But this just shifts the nouns in the original question. Why think that the human being cannot acquire non-natural properties? Baker supposes that the person possesses natural properties (though derivatively) and the animal possesses non-natural ones (though derivatively). Why not suppose instead that the person possesses both natural properties and non-natural ones *simpliciter*? The only reason for introducing the derivative—non-derivative distinction is because of an *initial* (and *undermotivated*) move to disaggregate the person and animal. Baker begins one of her more important articles asking the following: "What is the relation between, say, a lump of clay and a statue that it makes up, or between a red and white piece of metal and a stop sign, or *between a person and her body*?" (Baker, 2000, 144, emphasis added). The *starting point* is a disaggregated being. Another way to state the problem here is to focus on the mode of possessing a property. The person, on the constitution view, possesses natural properties but the mode is derivatively. On the hylomorphic view, the person possesses natural properties simply. Why *prefer* the former mode? Consider Hacker's five-dollar bill (2009); it is colored hazel and is worth five USD. A five-dollar bill possess each property simply. (This is a characteristic feature of substances, namely, their capacity to take on contraries). Which *mode* one chooses will be the result of a prior understanding of what persons are. The choice of which mode, then, does not argue for a view on what persons are. Preferring the former mode follows from an initial disaggregation of person and animal. And this initial disaggregation is not motivated by considering solutions to various metaphysical puzzles, since the Constitution view carries intuitive costs as well (see Burke (1994), Eberl (2006, 2020), Olson (2002)). I would add here that its solution to certain puzzles are illusory in the sense that they are considered *puzzles* only if one makes assumptions that are at home with the Constitution view, namely, disaggregating person and animal and then asking how we should understand their relation. So, the view starts with an initial disaggregation that is itself unmotivated.

usage according to which transitive "…consciousness is consciousness *of something*" (2013, 20). It is consciousness of or about something else. Transitive consciousness may be most synonymous with the notion of awareness. Intransitive consciousness, on the other hand, is consciousness "…one may lose (on fainting, when having a high fever, or being knocked out) and regain (on recovering consciousness)?" (2013, 19). It is contrasted with being unconscious, suggesting that intransitive consciousness is a power or capacity. For intransitive consciousness, one *has* consciousness; for transitive consciousness, one is conscious *of* or *about* something.[18] One might state the distinctions here more simply as between consciousness understood as a power, and consciousness understood as a specific state of an agent—and even more specifically, an epistemic state as in being aware of, or perceptual consciousness of. Understood as such, we have no reason yet for thinking that consciousness can be understood as a separate entity within or constituted by the human being.

A slightly clearer statement of the view that we are selves comes from Strawson (2003). He says that the human sense of the self is,

> the sense that people have of themselves as being, specifically, a mental presence; a mental someone; a single mental thing that is a conscious subject of experience, that has a certain character or personality, and that is in some sense distinct from all its particular experiences, thoughts, and so on, and indeed from all other things" (Strawson, 2003, 338).

Summarizing these features he states that the mental self is "experienced as, a thing… A mental thing… A single thing… Ontically distinct from all other things… A subject of experience… An agent… A thing that has a certain character or personality" (2003, 339).

The reason this is only a slight advance is because Strawson says that the self is experienced *as* a mental thing but it is also the "subject" of experience. The self, the one doing the experiencing, cannot be both the subject and object of experience. I cannot experience my *self* because I cannot be both object and subject at the same time. Again, Hacker is instructive. "No one other than a philosopher would ever speak of being conscious of seeing, hearing, tasting or smelling something, as opposed to being conscious of what one saw, heard, tasted or smelled" (Hacker, 2013, 31). Hacker's point is to resist thinking that introspection is evidence of a self. To do so suggests a picture of a mental homunculus inside the human being. Rather, *the human being* thinks, perceives, introspects, and so on.

Strawson also claims that "the self's claim to thinghood is thought of as sufficiently grounded in its mental nature alone" (Strawson, 2003, 345). We are not told exactly what its mental nature alone is. Presumably, the claim is that the mental thing is purely mental. It is ontologically distinct from the brain states that might cause it. However, the properties or characteristics that Strawson attributes to the "mental self" can all be said of the human being. The human being is a thing, it is a mental thing in some sense, and it is a single thing considered both synchronically

---

[18] Armstrong demurs (1984), but his example of intransitivity is that of an agent who sees a horse. The example is an example of transitive consciousness.

and diachronically, it is ontologically distinct from other things, it is a subject of experience, and an agent. All these properties can be ascribed to the human being. Snowdon's reflections on our self-conception are complementary here. Our self conception includes, he says,

> an idea of our origin and when we were born,… of our spatial temporal history since then,… of certain physical properties… of certain psychological and mental properties we have and have had, for example, or psychological capacities, the ability to see and remember, certain experiences, beliefs, and a personality…" (Snowdon, 2014, 83).

And when we think about what entity it is that has these things, "we would ascribe all these properties *to the human animal*… it has the same origins, history, material and psychological properties,…"(Snowdon, 2014, 83, emphasis added). We can more plausibly suppose that the human being is the subject of first-person mental states than to suppose some mental homunculus within the human being that is the true bearer of these states.

Hume makes a similar observation stating that, "when I enter most intimately into what I call myself, I always stumble on some particular perception or other … I can never catch myself without a perception, and can never observe anything but the perception" (Hume 1978, 252). Hume erroneously inferred from this that his self is a disunified collection of perceptions. It is equally consistent with his reflections, and with Snowdon's and Hacker's, that the so-called subject of consciousness is just the human being. This human being, you, and me, is a spatio-temporal continuant that has both mental and corporeal properties. Persons, on this view, are not bodies, but they are also not minds either. Persons have bodies, and they have minds. But a person cannot be identical to what the person possesses.

None of these reflections fall prey to Shoemaker's quip that when I look in the mirror I do not see my self. Whereas it is true that when I look at the mirror I do not see my self, understanding self as referring to a strictly mental homunculus in my human being. But understood in that sense, there is no self. What I do see when I look in the mirror is me.

Let's entertain for the moment what an exclusively mental view of the self would entail. It seems to imply that there is this mental substance ontologically distinct from any corporeal property, or corporeal being. The human animal cannot think; it cannot, on this proposal, be a self. The idea that I, qua person or self, have corporeal properties is actually a categorical error on this view. On one understanding of Strawson's view, if I'm identical to this mental self, I cannot have corporeal properties since the self is unitary, for Strawson, "in its mental nature alone" (2003, 345). I cannot be 6 feet tall, 185 pounds; I do not walk or run, I do not eat, or digest food etc. When I look in the mirror I do not see me. But if this is what the view entails, it doesn't have much by way of plausibility to it.

The self, understood as a strictly mental thing, seems to go in and out of existence. Its criteria for persistence seems to be self-consciousness, namely actual states of consciousness. But Strawson is clear that the self is not a power, or property. It is not the capacity for consciousness that is the self, but rather it is the thing that is conscious—it is the *subject* of experience. Understood in this way, the view avoids gappy existence (Forbes, 1985). But with such clarifications, it still looks rather plausible to say that the subject of experience is the *human being*—the physical, spatio-temporal continuant with a rational nature. At least, we are not given any reason for *preferring* a view of the self as a strictly psychological *entity* distinct from the human being.

From the proposition that the self is experienced *as* a mental thing, it doesn't follow that it *is* not a physical spatiotemporal continuant. A putative religious experience is experienced as being of God but naturalists no doubt will not accept the inference from being experienced as, to reality. Consider our experience of putative moral truths. Moral truths might be experienced as universally binding. The relativists or error theorists will claim that the content of such experiences are not indicative of reality. Furthermore, I might be aware of myself thinking through an argument; but I can also be aware of *me* lifting 175 pounds above my chest—an act that only an animal can do. I have mental properties, to be sure. But it doesn't follow that the thing that has those properties is itself mental *alone*, or that there is a *separate* mental thing in this human body. Again, a five dollar bill can have the property of being valued at five USD, and the property of being colored green and hazel. The subject of those properties, though, *is* the five-dollar bill (Hacker, 2009).

The clearest exposition of the view that you and I are a self, understood as a mental or psychological entity of some sort, is given by E.J. Lowe. He defines a self as, "a being that can identify itself as the necessarily unique subject of certain thoughts and experiences and as the necessarily unique agent of certain actions" (Lowe, 1991, 82). Tooley provides a similar definition of the person qua, holder of a right to life. "An organism possesses a serious right to life only if it *possesses the concept of a self* as a continuing subject of experiences and other mental states, and *believes that it is itself such a continuing entity*" (Tooley, 1972, 44, emphasis added).

Though these views are clearly described, they are not plausible accounts of persons. What makes these views implausible is the suggestion that the self exists only if one performs an epistemic act. For Lowe, it is forming a belief with a modal operator. One must be able to "identify" oneself "…as the *necessarily* unique agent of certain actions." And Tooley requires forming the belief that one is a "continuing subject of experiences."[19] If you and I are selves on such an account, you and I do not exist until we form such beliefs. Both beliefs require fairly developed intellectual capacities, ones that even toddlers or the mentally disabled might not possess (i.e., beliefs with modal operators). So, although Lowe and Tooley give clear expressions to the view that you and I are selves, such views define personhood much too narrowly.[20]

---

[19] One might add to this list Baker who requires that a person "*conceive* of herself as the subject of…thoughts" (Baker, 2005, 29, emphasis added). Here too, an epistemic act is required to be a person. The absurdity of the view can be illustrated by asking a simple question: who is it that does the conceiving? If we say the person, then persons cannot be identified with an epistemic act of conceiving of oneself, for the person already existed to do the conceiving.

[20] I think these views are plausible vis-à-vis giving an account of the self so long as one's notion of 'self' is not reified but refers instead to a constellation of psychological states and abilities; the

4.5.2   *Reifying the Self Is Arbitrary*

Sometimes, views like Tooley's (1983) and Warren's (1992) are criticized for being arbitrary. I think such criticisms are correct, but they need to be made with some thoroughness. Every viewpoint has intuitive starting points, but that alone is not enough to count as arbitrary.

Michael Almeida (2004) entertains an interesting case that motivates why such views are arbitrary. He has us imagine a community existing of beings who possess a standard list of capacities we associate with personhood such as rationality, intentionality, self consciousness, and so on. Call these standard human beings. Let's suppose that there exists an additional community that is, we may suppose, morphologically identical to the standard community. This community has the standard list of capacities but they also possess a capacity for "telepathy, psychokinesis, psycho projection, mental healing, and so on" (2004, 30). Suppose that all standard adult human beings lack such capacities. Call members of this latter community the advanced. Members of this advanced community might reason as follows: These advanced psychic properties are morally relevant for identifying those who have moral status. In fact, these properties are essential for identifying those who are full-fledged persons. Therefore, the standard human beings do not bear equal moral status.

The standard human beings might cry foul at such reasoning, but what exactly is the problem? The advanced can point to a principled distinction between them and the standard human beings. "We [the standard human beings] might simply insist… that normal adult human beings are members of the moral community and discontinue the discussion. Or we might charge the advanced community with an underdeveloped moral sense and end discussion in this way" (Almeida, 2004, 30). But these responses clearly do not give those in the advanced community any reason to change their criteria for having moral status. Almeida comments from the perspective of being a standard that "these responses simply beg the question in our favor" (Almeida, 2004, 30). We (those in the standard community) might object that the advanced are *simply asserting* that telepathy and the like are *necessary* conditions for having moral status. But how could we mount a response? We can point to the value of being conscious or having interests per se, without distinguishing between

---

term self is a *façon de parler* for psychological predicates (Hacker, 2009). So long as no entity or subject is implied, it is innocuous to state that believing oneself to be a continuant or as 'necessarily unique' is an important and highly developed psychological achievement. These views are not plausible if they intend to capture our intuitions on what *you and I are essentially*. The latter discourse must consider commonsense judgments to the effect that you and I were infants, toddlers, juveniles, and so on. It should also consider other judgments such as, "[w]e *seem* to be animals. When you eat or sleep or talk, a human animal eats, sleeps, or talks. When you look in a mirror, an animal looks back at you" (Olson, 2007, 23); and "we are physically identical with animals. We have the same developmental and evolutionary history as those animals have…How could things like that—being that no biologist could ever distinguish from animals—not be animals?" (Olson, 2007, 62–63).

being conscious as a standard vs. an advanced. But that meets one assertion with another. The advanced community can respond quite consistently by observing that the interests that matter are those who have the capacities for telepathy and the like. If having the capacities for telepathy and the like are necessary conditions for being full-fledged members of the moral community, then the interests and desires of only those in the advanced community matter. Again, the response "simply begs the question in our favor." The advanced community has drawn a principled distinction, but we can see clearly that it is viciously arbitrary.

Procedurally speaking, however, the advanced community is doing the same thing that those who make the exercise of consciousness a necessary condition for having moral status. Why draw the moral or ontological boundary there? One can distinguish among exercising a capacity well or poorly, actually exercising a capacity, possessing an immediately exercisable capacity (that awaits a decision to exercise it), possessing a readily exercisable capacity (such as, once impediments are removed—the person wakes up, the fog of alcohol wears off, and so forth), and possessing a remote capacity (such as, the ability to learn Icelandic after a lot of effort), and possessing a very remote capacity (such as, the zygote's ability to grow a brain and body that can learn Icelandic). To insist that any of these capacities except the last divides the human person from the human non-person is arbitrary. The point here is that if one draws a moral boundary between the exercise of consciousness and being the kind of thing that develops the capacity for consciousness, that is an arbitrary drawing. For the function view, only things evincing the former abilities are persons. One might think that drawing such a line is principled and might even agree with it. So too do members of the advanced community agree with their criterion.

One might think that the BT experiments, or the thought experiments entertained in the last chapter from Wellman and Puccetti were meant to argue that such criteria are not arbitrary. I have addressed part of this claim when assessing such thought experiments. We can consider extensions of the current thought experiment to see that our intuitions go in a different direction: we value *capacities* but we also value the *things* who develop those capacities.

Consider the advanced community again except that you are one of them and that you are a true believer in its criteria for demarcating the moral community. Consider a permutation of a burning IVF clinic scenario according to which you are a first responder tasked with saving one of two embryos—suppose you cannot save both. You know that one embryo was created by standard parents, the other by advanced parents. Assume, per actual world, the facts of human development, specifically, that healthy human embryos in a supportive gestational environment develop along a continuous and self-directed trajectory. Assume also that you know of an organization that manages embryo adoptions for advanced humans (a feature similar to our actual world). Whom do you save? You would pick the advanced embryo. But this choice suggests that you value members of the advanced community *at any developmental stage*. Actual displays of telepathy and the like are not necessary. What is necessary is being the kind of being that develops such valuable abilities.

So even from the perspective of a member of the advanced community, the kind of being one is morally matters.[21]

Two conclusions follow from this brief comment on arbitrariness. First, it is arbitrary in a bad way to distinguish between beings that exercise consciousness and being the kind of thing that develops consciousness. Second, our moral intuitions 'track' valuing things that develop such capacities, we value things of a certain sort.

4.6 The Problem of Underdetermination Again

The key defect throughout the discourse is underdetermination. The underdetermination comes in two sorts. First, we discovered that even if the intuition we are supposed to have in response to the BT-experiments is correct, it still does not follow that you and I cannot be killed by an early abortion. Second, the intuitions we have concerning selves and the identity of the self do not disambiguate between a self understood as reified, versus a self understood as a *façon de parler* for one's psychological states and abilities (Hacker, 2009). We have no reason for thinking that you and I are identical to consciousness versus having consciousness.

The problems here can be summarized as follows. Suppose we understand self-consciousness as reified into a separate substance all its own. But self-consciousness can be understood either as a state, or a capacity. I cannot, however, be identical to either a state, or a capacity. I experience states, I might find myself in a particular state of mind. But I cannot be identical to a particular state; I cannot be identical to what it is I experience or to that by which I experience it. No thing can be identical to one of its parts. Likewise, I have a capacity for self-consciousness, but I cannot be identical to the capacities that I have. Change the language slightly and the claims ring true as well. I have the power of self-consciousness, but I am not identical to it; I have the ability to reflect on my own conscious states, but I'm not identical to that ability. If I were just that ability, I would not be able to run, walk, eat, sleep and so on. Self-consciousness is a power that I have, but not what I am. Again, the difference between the hylomorphic view and a psychological view is the difference between being the kind of *thing* that develops and manifests an array of rational and psychological abilities, and being only that thing that *exercises* an array of rational and psychological abilities. It's the difference between being a certain sort of *entity* and being an entity that *exercises* certain sorts of prototypical powers. Once we isolate out the primary distinction, we can see better that none of the typical thought experiments used to motivate the latter view serve as exclusive motivations for it. The former view can grant that, for example, self-consciousness is a very important

---

[21] The reader is welcome to fill in further details as is necessary. Change the case to a choice between saving a advanced embryo and a monkey embryo. Or, add that you also know personal female members of the standard and advanced communities willing to adopt an embryo. However the details are filled in, you still choose the advanced embryo. And that choice is enough to infer that you value the kind of thing that embryo is.

and even essential ability of the human person. A thing that held out no possibility for developing and realizing the capacity for self-consciousness would likely not count as a human person even on the former view. But claiming that it is not just important but required, that it must be *actually realized* or *exercisable* is under-motivated by any of the data proponents of the psychological account provide.

Even if it is the case that the functional view is coherent, we do not have an argument or reason *for* (pk) because we still have no reason for thinking that persons do not come into existence *until* psychological capacities are functional.

To summarize the previous two chapters, one can see that the de jure defects are parasitic on what is presumed by one's interlocutors. For the TRIA argument, what is presumed for the proponent of the TRIA is that killing is inconsequential. The defect is that the argument presumes (pk). Since a justification for (pk) is what is asked for, the de jure defect here is serious and poignant. This might best be catego-rized as an inference-defect because circular arguments do not provide a reason for a claim (see Sect. 3.4). The de jure defect for the no-person strategy is that the view of personhood required to make good on the argument is undermotivated by the typical reasons offered for it. In the absence of a motivation for the function view, (pk) remains unsupported. This is an iteration of a justification defect satisfying condition (ii).

But the de jure defect here is not as poignant as that affecting the TRIA. The reason is that (pk) itself is not assumed by those who argue for an embodied mind or constitution view of the person. What is presumed are understandings of what person and body mean, or what person and animal or organism mean—(pk) *itself* is not presumed. As noted above, Shoemaker's remarks suggest that by person is meant you and me, and by body is meant, everything else; heart, lungs, kidneys, and so on. These assumptions are not linked to (pk), as far as I can see. But they are unmotivated assumptions. So, the de jure defect here is a serious defect, but not as serious as that affecting the TRIA argument.

Chapter 5
The Modal Argument Against Abortion

5.1 Introduction

The previous chapter discussed arguments for the permissible killing of unborn human beings (pk) that focused on arguing that what is killed is not a person. If there is no one there to be killed, there is no one who can be harmed either. The no person strategy, as I have called it, relies on arguments for a broadly psychological account of the human person.

The previous chapter argued that once we focus attention on the motivations for one such psychological account, namely the embodied mind account, we find that such motivations do not exclusively support a psychological view of the person. The data or evidence of our intuitions on brain transplant thought experiments, for example, is not evidence that you and I do *not* exist when an abortion might be performed. Different views of the human person can accommodate the intuitions we have concerning brain transplant thought experiments, namely the hylomorphic/ Aristotelian view mentioned previously. The de jure defect is that the embodied mind account *underdetermines* a premise in the no-person argument, that premise being that the developing human being is *not* a person.

The chapter closed by considering a more direct argument to the effect that the developing human being is not a person, namely that you and I are selves understood as an enduring mental entity. This view suffers from de jure defects, but one that many philosophical views suffer from, namely, counter-intuitive consequences. When we understand the self as a thing, that you and I are identical to this self, and this self is a mental thing alone, we do not have corporeal properties. If the view admits that you and I can have corporeal properties, it is then compatible with the view that the human being has the relevant mental properties. Once we pollard away an understanding of the self as being a mental thing *alone*, the view does not motivate the no-person argument for (pk).

This chapter explores an argument that it is not permissible to intentionally kill pre-born human beings, i.e., ~(pk). I explore the extent to which such an argument

S. Napier, *Justified Killing*, https://doi.org/10.1007/978-3-032-14946-6_5

suffers from any de jure defects. I observe that the defect suffered by the present pro-life argument is neither that which the no person strategy nor the TRIA argument suffers. The defect it suffers from, if it is a defect, is that it assumes the human person has worth *in itself*.[1] The human person is precious or worthy even if she or he cannot engage in so-called worthwhile activities. So, the argument here argues that such a being exists when abortions may take place.

5.2   The Argument for ~(Pk)

The procedure in this section is somewhat unorthodox. Because one issue in the abortion debate concerns what is it that is killed, one often sees arguments for and against the personhood of the unborn human being. The strategy for most arguments for ~(pk) is to motivate a metaphysical view of the person and then apply that view to unborn human beings. The idea behind the strategy is sensible from both a philosophical standpoint and from the standpoint of basic moral psychology. One applies without prejudice a plausible metaphysical view of the person and if that view entails that what is killed is you or me, the moral conclusion follows. One should follow the argument where it leads.

The argument presented here starts instead with our intuitions on harm. The reason for doing so is because both proponents of (pk) and ~ (pk) have the same intuitions on impairment arguments discussed in Chap. 3. And those arguments are based on intuitions we have regarding harm, even prenatal harm. The modal argument presented here is a permutation of prenatal harm cases—it represents the limit case for prenatal harm cases.

The basic idea is to consider an actual world case where someone improbably survives an abortion. Once we pair our intuitions on that case with accepting Transworld identity (TWI) it follows that abortion would have killed a person. TWI is the view that you and I exist in other possible worlds. Consider the counterfactual that you could have gone to a different college, or high school, or chosen a different career path. It would still be you who goes to that different college, high school, or engages in a different life plan. Propositions of the form "I could have chosen to … " or "I could have experienced …" can be true, depending upon how one fills in the blank. If they can be true, we exist in different possible worlds. If one considers cases in which someone survives an abortion and grows up into adulthood, call her Sarah, the intuition is that Sarah was almost killed. (Being killed is the terminus to a prenatal harm case.) Since Sarah was chosen at random, anyone of us could have been almost killed by an abortion. It would follow that the abortion almost killed us. In those worlds in which the abortion was "successful," it would have killed me, or

---

[1] For why this is a safe assumption, see Napier, 2020, ch. 5; Debes, 2009; Kaczor, 2013; Sulmasy, 2008; Kass, 2008, and Zuniga, 2004.

you, or Sarah, etc. It follows that abortion kills a person. I refer to this as the modal argument against abortion.

5.2.1   The Modal Argument Against Abortion

Assume that events in this world (the actual world) could have gone otherwise. Things could have been different than they are. Human agents could have chosen different activities than they do; events outside of human agency (if any) could have been different than they are and so on. These thoughts suggest the possibility of transworld identity TWI. Introducing the notion of TWI, Alvin Plantinga says the following,

> It is natural enough to suppose that the same individual exists in various different states of affairs. There is, for example, the state of affairs consisting in *Socrates' being a carpenter*; this state of affairs is possible but does not in fact obtain. It is natural to suppose, however, that if it *had* obtained, then Socrates would have existed and would have been a carpenter; one plausibly supposes it *impossible* that this state of affairs obtain and Socrates fail to exist. If so, however, then Socrates *exists in* this state of affairs. But of course, if he exists in this state of affairs, then he exists in every possible world including it. For clearly every possible world including *Socrates' being a carpenter* also includes *Socrates' existing*; each such world is such that had it been actual, Socrates would have existed. So, Socrates exists in many possible worlds (Plantinga, 1974, 88).

What is true for Socrates in this passage is true for all of us.

Consider the point on analogy with books in a library. Each book represents a possible world (suppose these books are really thick, and that there is an infinite amount of these books representing an infinite number of possible worlds). Consider just two of these books, each of which includes details about your life. Suppose one of these books represents the actual world and the one next to it represents a possible world whose 'chapters' are identical all the way up to this present moment. In the actual world, you choose to go skiing in Big Sky, in the possible world you do not. You *could* have skied elsewhere or not skied at all; but if you did, it would have been *you* who does these other activities. Of course, different 'books' with you in it could represent all the different ways your life could have gone, but you exist in these books. You exist in different possible worlds by which is meant, if that other world had been actual, *you* would exist. This is the basic idea with TWI.

Consider the same idea visually: consider the actual world up to the very present moment as if along a line. Suppose that the line forks into two directions. The fork represents a choice: You have the choice to put this book down and stop reading or to continue reading. Suppose the actual world is the one in which you continue reading. You *could* have, however, stopped reading. And if you would have, it would have been *you* who stops reading. These choices represent your life going in different directions along the two lines. Of course, the lines might be represented as being close together, noting that your choice to continue reading is followed by the exact same activities that you would have done had you stopped reading. Also imagine the complicated number of lines and their various branches for each moment of one's life corresponding to the numerous choices and events that could have gone differently. It is still you, or me who experiences or chooses differently. That is as easy to suppose and accept as anything can be accepted in philosophical discourse.

Plantinga acknowledges that certain philosophers object to TWI, but there are also philosophers who object to the existence of the external world, the existence of medium-sized objects, global skepticism, and so on. There may be elaborate objections to such positions, but the intuitive plausibility of TWI remains.[2]

With the basic idea of TWI in hand, consider an abortion survivor in the actual world, named Sarah Smith.[3] She retells her story as follows:

> Twenty-nine years ago [as of 1999] my mother decided to have an abortion. At the time, she was pregnant with twins, but nobody knew this, not even her doctor ... It's frightening to think I was almost aborted when my mom had a D&C abortion. Somehow, miraculously, I survived! My twin brother wasn't so lucky ... Several weeks later, my mother was shocked to feel me kicking in her womb. She already had five children and she knew what it felt like when a baby kicked in the womb .... She went back to the doctor and told him she was still pregnant ... and that she wanted to keep this baby .... After surviving the abortion, I was born with bilateral, congenital dislocated hips and many other physical handicaps .... At six weeks I was put into my first body cast. Many surgeries and body cast followed over the next few years (Smith, 2012)

Sarah survived an abortion by accident. She got lucky—if you think that not getting killed is a good thing. The abortion doctor failed to do something he routinely does which results in Sarah surviving, but it still permanently injures Sarah. (Another abortion survivor, Claire Culwell, reports that she was born with a dislocated hip, club feet and suffered numerous issues as an infant due to the D&E abortion that was performed on her identical twin (LaPointe, 2014).) Sarah survives, is born, and matures in a way typical of human beings. We can easily imagine that things could have gone differently than they did. In most every nearby possible world in which Sarah exists, she is killed.

If one doesn't agree yet that it is *her* that is killed, consider first an abortion performed late in gestation in world W_i, for example at 22 weeks gestation. On McMahan's account, what is killed is a person. Now imagine the abortion is attempted in progressively earlier stages of pregnancy in an iterated series of possible worlds (symbolized as W_{i-n} where $n \geq 1$). Suppose also, however, that for each abortion attempted in progressively earlier stages, there is a corollary world in which she survives—suppose that she survives in the actual world W_a. As noted, it is plausible to say that since Sarah is killed in W_i, she is also killed in W_{i-n}. Using the

---

[2] For more discussion see Penelope Mackie, "Transworld Identity", *The Stanford Encyclopedia of Philosophy (Fall 2008 Edition)*, Edward N. Zalta (ed.), URL = http://plato.stanford.edu/archives/fall2008/entries/identity-transworld/

[3] There are numerous other actual cases such as Nic Hoot, https://www.youtube.com/watch?v=-9eMIexNNo4,whowas born alive after a failed abortion attempt. He lost one leg below the knee, and the other above the knee as well as deformed hands. Another example is the story of Gianna Jessen, http://www.giannajessen.com/EPK/bio.html,who suffered cerebral palsy as a result of the attempted abortion.

actual world as our reference point, consider other nearby possible worlds in which the abortion happened to be performed at an earlier stage of development. When Sarah in W_α considers what could've happened in $W_{i\text{-}n}$, she is still entitled to say that she was almost killed. From the first-person perspective, this is the most natural thing to say. Had the abortionist not made an uncharacteristic 'error' Sarah would have been killed.[4]

Of course, one can resist this way of framing the argument. All parties should agree that it is correct for Sarah to say that had it not been for this uncharacteristic 'error' (whatever that error may be), she would *not have existed*. To say that one would not have existed in $W_{i\text{-}n}$, is different, however, from saying that she is killed in $W_{i\text{-}n}$. And what the modal argument needs is to make plausible the claim that Sarah is killed in those worlds.

In response, the fact that Sarah would not have existed is not itself a reason for thinking that Transworld identity does not hold between W_α and $W_{i\text{-}n}$. Sarah in the actual world very narrowly escapes a botched abortion which leaves her with life-long injuries, and facial disfigurement. If the abortion did not injure her, who did it injure? It does not seem to matter how early in the pregnancy those injuries were inflicted on her. From the first-person perspective, Sarah could plausibly say "what the abortionist did to *me* caused these injuries."

I realize full well that something like a functional view of the person, or some permutation of a body-self dualism entails that these assertions are false (or that Sarah would be understandably confused that it was *her* who suffered such injuries). And such views of the person do not impugn Transworld identity, since Transworld identity holds between psychological entities too.

Such theories, however, do not provide a natural interpretation of abortion-survivor cases. Moreover, to think that mentioning such theories stymies the modal argument would misconstrue the dialectic to this point. First, the previous chapter argued that key psychological views of the person are underdetermined by the typical thought experiments used to motivate them; and they do not entail a view about when persons come into existence. Of course, such views can be accepted provisionally to function as putative objections to the modal argument. Given the preceding work and following philosophical convention, the initial statement of the modal argument can engage in some deferred maintenance, as they say, and address psychological views as objections below. But I have already shown them to be underdetermined.

Second, the modal argument is framed deliberately to mirror prenatal impairment cases the basic intuition of which my interlocutors accept. When *an abortion* survivor entertains what happened, the natural and commonsensical response is for her to say that she was almost killed. As an adult, Sarah can say that the suction catheter used to suction out her brother exerted enough force to dislocate her hips. Sarah was injured by the abortion. When did this injury occur? The obvious answer

---

[4]This is not a sorites argument. A sorites argument may occur when discussing the identity of quantitative nouns, such as heaps. Here we are identifying our intuitions on when a killing occurs.

is "when the abortion occurred." So, Sarah existed when the abortion occurred. My interlocutors should not find this response surprising.

So, Sarah can truthfully say in the actual world,

(P): I was almost killed.

The term 'almost' should be read as a counterfactual as follows,

(P'): I could have been killed.

And the counterfactual can be rendered as,

(P''): Had W_i (or W_{i-n}) been actual, I would have been killed by an abortion (A).

This thought experiment is meant to show that (P'') is true. And for (P'') to be true, Sarah in W_a (the actual world) = Sarah in W_i (a world in which she is killed). To say truthfully that "*I* could have been killed" is to make a counterfactual claim and it entails that Sarah exists in that world, until Sarah was killed. There is reason to suppose that identity holds between Sarah in W_a and Sarah in other abortion-successful worlds.

At this point, the argument gets technical, but easy enough to follow if we keep in mind that possible worlds are *logically* possible worlds. What happens to Sarah can happen to any one of us in some world or other. There is a dyad of possible worlds in which for any adult human being S, S narrowly escapes A in W_s but is killed by A in W_k. Let s in W_s and k in W_k be variables for all and only those worlds in which for any S, S implausibly survives A in W_s, but is killed by A in W_k. W_k is identical to W_s in all respects except that for any S, S does not survive A in W_k. If for any S, S can *truthfully* utter P'' in W_s, then S is killed in W_k. If TWI holds and the referent of 'I' in (P'') is a person, a person is killed by A in W_k. So,

(Id): $\square$ (S in W_s = S in W_k).

Since identity relations are necessary and S is a person in W_s, it is necessarily true that if S is killed by A, A kills a person.

(kp): $\square$ (S) (S is killed by A $\rightarrow^5$ A kills a person).

I take it that the truth of (kp) provides sufficient justification for thinking that abortion is intrinsically immoral.[6]

The intuitive idea behind this argument is that we can run a structurally similar thought experiment on everyone as we did for Sarah. It is broadly logically possible that you and I exist in a world in which we narrowly escape an abortion and the nearest possible world to that one is identical in all respects to the world in which we survive except that we do not survive. If TWI holds, what is killed in W_k is a person, namely, S. (I am supposing there is no objection to whether S exists in W_s,

---

[5] $\rightarrow$ should be interpreted as a conditional here not as denoting logical entailment.

[6] In taking (kp) sufficient to prove that abortion is immoral, I am assuming the conclusions drawn in Chap. 3. The TRIA, as discussed, may grant that you and I may be killed by an abortion, but it argues that it does not harm or wrong us in that there is nothing of value that is destroyed.

but there is, as noted, an objection to supposing that S exists in W_k an objection that is addressed below).

The thought experiment begins by considering abortion survivors and what they would be justified in saying about the abortion attempted on them. Considering such cases gives (P″) significant credence, and (P″) gives one considerable reason to accept (Id). (Id) paired with the premise that S is killed by A, entails (kp). Since I am supposing that such a world-dyad (W_s and W_k) is broadly logically possible, there is at least one possible world, i.e., W_k, in which S is killed by A. More formally,

(M): $\Diamond$ (S)(S is killed by A).[7]

I take (M) to be quite plausible, for it is logically possible that a person is killed by an abortion. I also think that (P″) is quite plausible. I might even suppose that (M) is obviously true on modal grounds in that there is no obvious conceptual or categorical truth that would impugn the truth of (M). It is broadly logically possible that for any one of us, we could have been killed by an abortion. Refuting the present argument requires refuting what appears to be an obvious logical possibility. There are, however, reasons for thinking that (M) is false and with it, (Id) is rendered false as well since (Id) entails (M). And if (Id) is refuted, then the inference to (kp) is broken and the argument fails via *modus tollens*. I entertain such an objection further below (Sect. 5.3), but first an explanation for why this argument is important.

5.2.2   Why the Modal Argument?

The reason for starting the case for ~(pk) at this point is that it relies on intuitions that proponents of (pk) would accept. As noted above, the modal argument extends a case of prenatal harm the intuitions of which are shared by proponents of (pk) and ~ (pk). For example, McMahan supposes that a pregnant woman ingests a drug that slightly alters the mother's mood but it is known to render the developing child sterile. "Virtually everyone agrees that the pregnant woman's action is seriously wrong" (McMahan, 2002, 280).[8] Everyone agrees, that is, that in cases in which the child survives, the child is harmed *by* the prenatal intervention. It appears, then, that proponents of (pk) should grant a key claim in the modal argument, namely, Sarah can truthfully say in worlds in which she survives that "I was almost killed."

---

[7] For modal logic buffs the argument assumes S5 (Konyndyk, 1986). (M) entails $\Box\Diamond$(S) (S is killed in A), which reads: "it is necessarily the case that it is possible that for any S, S is killed by an abortion." The reason is that what is possible entails that it is true in some world, say, S is killed in W_i. If that is the case, it is true in every other world that 'S is killed in W_i'. So, necessarily, it is possible that S is killed by an abortion. Since suffering an abortion is logically possible, it is true in the actual world, that you and I are killed by an abortion in some possible world. And this comes to the same conclusion reached in the text, abortion kills persons.

[8] See too DeGrazia's discussion of prenatal harm cases 2005, 289 ff. and Chap. 3.

One might think that the modal argument is circular on this point, circular in a similar way that the future of value argument is circular—as is explained in Sect. 1.4. The future of value argument must assume that the thing that has a future is ontologically identical to you or me. Likewise, asserting that it was *Sarah* who was almost killed assumes, but does not argue for, Sarah existing at a time when abortions are typically performed.

This riposte misunderstands how the modal argument proceeds. When we consider a prenatal harm case in which the child survives, we all have the intuition that the child was harmed by the prenatal intervention. One could say that the cause of harm and the harm can be disaggregated (in terms of occurring on different *entities*), but this would only be plausible in very limited circumstances discussed below. In general, what causes the harm is when the harm occurred. Philip Montague entertains an interesting case in which he develops a serum that when squirted on your skin renders you infertile without you knowing it. In squirting the serum on you, Montague harms you. This is true even if Montague sets a booby-trap that squirts this serum onto your skin next week. Montague comments that, "[E]ven though you are not sterilized until next week, the action by which I violate your right occurs not next week, but now as is clear from the fact that, even if I die immediately after setting my trap, I violate a right of yours nevertheless" (Montague, 1989, 68).

So, consider again one's intuitions on a prenatal harm case *in which the child survives*. Our intuitions are that the child is harmed by the prenatal intervention. Now consider prenatal harm cases in which the harm is extended to its terminus, namely, the child is killed. As argued in Chap. 3, to think that this is a morally neutral act must think that killing is a morally neutral act. That would already assume the truth of (pk). What is important to observe here is that proponents of (pk) or ~ (pk) have the same intuitions regarding prenatal harm cases in which the child *survives*. The intuition is that it is the *child* who is harmed *by the intervention*. The modal argument is merely extending the severity of the harm out, so to speak, to its terminus, namely, killing and adding a world dyad in which the child survives. The only twist for understanding the modal argument is that we are asked to see things from the first-person perspective of Sarah in W_s in which she survives. So, if we compare the actual world in which she survives with W_k and we suppose TWI, it follows that Sarah was killed in W_k. The life of the child is ended by the intervention, in this case an abortion. To repeat, we are to imagine these cases side by side in which the child survives in one, but is killed in the other. If we say in the survival world that Sarah survives the abortion, we have to say that it is *her* who does not survive in the abortion world.

Why is this procedure not circular? That Sarah (or the human being) does not survive in the abortion world is not itself a reason for thinking that Sarah *never* existed in that world. That is, suppose you want to play nicely and grant that in the abortion world, Sarah does not survive. Granting that much is compatible with either Sarah not ever existing in that world, or with Sarah being killed. Notice, however, that what is granted is not a reason for preferring the former. To see this more clearly, suppose we do not name this entity. Label the entity in the abortion world as entity X. That X *does not survive* in that world does not give one any reason for thinking that X is *not* Sarah; it gives one no reason to deny Transworld identity with Sarah. Suppose that X exists in both W_k and W_s (per assignments above where s = the survival world and k = the killed world). If we have the intuition that X = Sarah in W_s, X's death in W_k is clearly not a reason for thinking that X $\neq$ Sarah in W_k. (In fact, I find my intuitions going in the other direction, but I am trying to be fair with my critic.) Of course, one reason for thinking that X = Sarah in W_s is the intuitive pull of abortion-survivor cases, and one might deny this intuition *ab initio*. But that looks too convenient; convenient in a way similar to how some animalists deny the brain transplant intuition (as discussed in Sect. 4.4) or how some deny McMahan's fetal death intuition.[9]

So, the modal argument deflects charges of circularity as follows. The proponent of the modal argument can rely on the following observation that is decisive: *the ontological identity of what is harmed does not change when we kill it*. That we kill an entity is itself not a reason for thinking that Transworld identity does not hold between that entity and the same entity in worlds in which the entity survives.

The same point can be revealed if we view the modal argument along a continuum with prenatal harm cases. Again, everyone agrees that the child is harmed by the prenatal harm when the child survives. That the child is killed by a prenatal intervention, however, is not *itself* a reason for thinking that the identity of what is harmed changes or that what is harmed in the survival case doesn't exist in the world in which that same entity is killed. In general, the severity of harms is not itself a reason for thinking that the ontological identity of what is harmed has changed. As we progress through more and more serious harms to their terminus in killing, we do not have any reason for thinking that with each progressive stage the entity harmed has changed its identity.[10] That is why the modal argument does not require *assuming* that Sarah is killed. Rather, we have no reason for thinking that she does *not* exist; and every reason given our intuitions in the survival world, that she does exist at the time of the harm. Understood as such, the modal argument does not assume ~(pk).

So, even though the burden of proof is on the argument for ~(pk) in this chapter, the modal argument meets this burden because it begins with intuitions that are granted by proponents of (pk), namely, intuitions on prenatal harm cases or metaphysical intuitions on what count as good reasons for thinking that Transworld identity does not hold.

A second benefit of the modal argument is that its central intuition gives one *direct* reason for ~(pk). Accepting the key intuition, namely, that you and I could have been killed by an abortion, *entails* that what is killed in an abortion is a person—assuming that you and I are persons. In contrast, the brain transplant thought experiments do not logically entail that what is killed by an abortion is not a person. One would need to supplement one's intuitions on the brain transplant thought experiments with *explanations* for why we find ourselves having such intuitions.

---

[9] See McMahan (2002, 78) and Napier (2023).

[10] This is why I think that Hendricks (2019a, 2019b) is correct as against Blackshaw (2019).

But the explanations vary, and as argued in Chap. 4, only some mediate the inference to (pk).

A third benefit of arguing in the way that the modal argument proceeds is that the metaphysical intuitions required to appreciate its plausibility come down to two: Transworld identity, and that an abortion survivor can truly say "I was almost killed." The modal argument is metaphysically economical, at least in its opening. Engagement with objections has to descend to the specific metaphysical assumptions ensconced in those objections.

Those are the benefits of starting with the modal argument. Let's turn to consider how it handles objections.

5.3   Objections

There are basically two objections to the modal argument as stated: the first is that it assumes that it is wrong to kill persons. What exactly is the defect? Recall Boonin's riposte to the essence-property argument canvassed in Sect. 1.4 to the effect that the debate on abortion concerns whether you and I have a right to life from the moment we come into existence. Boonin's position is that the essence argument[11] is circular because it takes for granted that it is wrong to kill persons; it takes for granted that you and I have the same rights at every point at which we exist.

The second objection is that aside from the intuitions the modal argument invites us to have, they are simply wrong. Persons do not exist when typical abortions may be performed. This objection follows the same strategy as the TRIA response to the impairment argument. The strategy in response to the impairment argument is to argue that so long as the prenatal child dies, there is no time-relative interest that is violated. Likewise, the present objection aims to argue that so long as there is not yet a person, there is no one who is killed. If one thinks that harm or wrongdoing has something to do with interest, then if there is no possibility for having interests, there is no possibility for harm or wrongdoing. Likewise, if wrongdoing has something to do with being a person (understood psychologically), then if there is no possibility for being a person there is no possibility for wronging anyone.

This second objection grants that one can craft the modal argument without presupposing any specific account of the human person. One might have the intuition that in the survivor world, the survivor can say with considerable plausibility that "I

---

[11] Boonin presents the essential property argument as follows: "P1: I am the same individual living being as the zygote from which I developed. P2′: I am a person essentially. P3′: If an individual living being has an essential property at one point in time, then it has that property at every point in its existence. C: The zygote from which I developed was a person" (Boonin, 2003, 52). As is recorded in Sect. 1.4, Boonin takes issue with P2′ according to which the term person can mean either biological organism, or it can mean someone with a right to life. And he says of these options that "[t]he first claim needed to sustain the rights-based argument against abortion maintains that the fetus has a right to life, not merely that the fetus is biologically human" (2003, 52–53).

was almost killed." The survivor is, however, understandably confused. Fully informed of the literature on personal identity, she should say that "I almost *didn't come into existence.*" Abortion, on the psychological account, is ontologically equivalent to contraception. Specifically, proponents of a psychological account of personhood and personal identity will take issue with (M). As explained in the previous chapter, such an account says that persons are psychological entities of some sort. They must have *exercisable* psychological capacities. To function as an argument for the no-person strategy, such an account must include the claim that persons do not arrive on the ontological scene *until* a human organism can manifest or exercise psychological capacities.[12] So, during the time in the organism's development that most abortions take place, there is no person. This looks like a de facto objection, but as is observed in Chap. 2, this can function as a justification defect according to which counter evidence is available, namely an alternative account of persons than what is entailed by P″. And this alternative account aims to weaken the strength of the modal argument by offering counterbalancing reasons that resist its initial intuition.

5.4   Responses to the Objections: Identifying the de Jure Defects

5.4.1   Where or What Is the Burden of Proof?

Regarding the first objection, the modal argument falls prey to it. It is unclear, however, just how serious this defect is (Chappell, 2004). First, Boonin is correct to say that "there are many rights that I possess now that I did not possess in earlier stages of my development" (Boonin, 2003, 53). For example, a right to vote or a right to an education are both parasitic on being *able* to vote or be educated. But that is a rather jejune observation in this context. The issue is whether you have a right not to be intentionally killed, and *that* right does not appear to be developmentally indexed. In contrast, a right to vote requires having developed cognitive abilities. A right not to be intentionally eliminated is not similarly labile—think of one's right not to be a victim of violence, sexual assault, battery and so on. If I exist, I have a right not to be battered. Our right not to be the victim of violence is not parasitic on our developmental stage—even infants enjoy such rights.

Second, Boonin's procedure is typical in these debates and at this very juncture in the dialectic. His strategy is to bifurcate person into two different meanings, one of which is morally innocuous (what is so valuable about being biologically human?), and the other is morally important but it takes for granted what is at stake (why assume that you and I have always had the right to life?).

---

[12] See Hasker (1999, 171–203).

The problem with such a procedure is that it works both ways. Consider for the moment consciousness which, according to many, is a marker of personhood or a symptom of a thing's moral status. Consciousness can mean the occurrence of P 300 potentials which is morally innocuous (what can be so valuable about having P 300 potentials?). Consciousness can also mean an awareness of one's environment or one's self—if there is such a thing. This awareness might be morally important but why assume that it functions as a *necessary* condition for having moral status? (For a parallel, why assume that you and I existing is sufficient for having a right to life?) No one disagrees that consciousness is a valuable thing to have. Assuming, however, that it is a necessary condition begs the very question at stake. Moreover, there are numerous counterexamples to such a criterion such as Lee's (2010) temporarily comatose patient, Schwarz's (1990) Uncle Otto case, Beckwith's (2011) Uncle Jed case, and so on.

Consider also Boonin's claim that having a body entails a right that no one use it (involuntarily). Here too we can consider body in strictly organismic terms and ask what can be so special about it such that no one can use it? Or we can understand body in more morally charged terms such as it being my body, or the body of a person. It does not follow even from these descriptions that having a body discharges my parental duties to my children (see Sect. 3.7.1 for further comment). To assume that it does would assume exactly what's at stake. So, when we interrogate Boonin's own favorite moral claims in the way that he interrogates person, we get equally dismal results.

The first two observations here focus attention on the epistemic question of when an argument can stop at a basic belief? The final claim in the modal argument is that it is impermissible to kill you or me—it is impermissible to kill persons and persons exist when abortions can occur. Isn't this enough? May not this premise function as a basic belief?

To say that being a person is not enough to render an act of killing that person immoral requires assuming (pk). That claim is the default position; 'it is permissible to kill persons unless proven otherwise.' There must be a reason for *not* killing a person; a reason additional to being a person. So, though the modal argument assumes that it is wrong to kill persons, the opposite view makes what many might think is a gratuitous assumption about the permissibility of killing persons. Put another way, the modal argument does not meet the burden of proof *if* that burden assumes that being a person is not enough to render killing that person unjust. The reason this defect is weak, however, is because it is questionable whether such a burden needs to be surmounted.

It is fair in philosophical discourse to require further support for premises in an argument if for no other reason than one is trying to find the principle or foundational basic belief. If one needs to know *more* than that S is a person before thinking she is unjustified in killing him, one must think that being a person is not enough to think that killing him is unjustified. And thinking that it is not enough entails that killing persons (knowing nothing else) is permissible (cf. Hershenov, 2023). I see no reason why such a position should set the burden of proof against the proponent of the modal argument. My interlocutors grant that born infants or children should not be intentionally killed because they are persons. Being a person *is* enough to explain why it is impermissible to kill them.

There is an additional reason for why the 'person is not enough' claim should not set the burden of proof: First, the chief reason for thinking that it is permissible to kill persons was addressed in Chap. 3, namely, the time relative interest account of harm and wrongdoing. We discovered there that the TRIA argument for (pk) is viciously circular in that it fails to provide a reason for (pk). A more promising approach was addressed in Chap. 4 according to which what is killed in an abortion is not a person and therefore no one is harmed. To the proponent of the no-person strategy, the modal argument is addressed and therefore can legitimately take for granted that it is impermissible to kill persons. Given the conclusions reached in Chap. 3, there should not be any proponents of the TRIA argument. It is a dialectically inert position.

5.4.2   *Psychological Theories and the Self Response*

The second objection (hereafter the self response) claims that we (persons or selves) cannot exist when an abortion typically occurs. The reason why is because personhood is a categorical claim according to which persons are not the kinds of things that can be killed by an abortion (at least not when most abortions may take place). Since the reply is making a categorical claim, it is logically impossible to kill a person due to an abortion. So, this view of the self is committed to the denial of (M).

$\sim$(M) = $\sim\Diamond$ (S is killed by A),

which is logically equivalent to,

$\Box$ $\sim$ (S is killed by A).

Just as human beings cannot have the property of *being a chicken*, so persons cannot have the property of *being the victim of an abortion* (assume that all and only abortions take place prior to the onset of consciousness). Label this property V (for victim of an abortion), and its complement (i.e., *not* having the property of *being the victim of an abortion*) as the complement of V, or comp-V. Since every person, qua person, has comp-V, it is an essential property of being a person. I understand an essential property to be defined as follows: "A property P is an **essential** property of an object x iff x could not exist and lack P, that is, ... iff x has P at every world at which x exists" (Forbes, 1986, 3).[13]

---

[13] This reply is very similar in spirit to Olson's (1997, 79ff.) when he is criticizing the psychological account. The difference is this: Olson focuses on the counter-intuitive consequence that on the psychological account, the fetus must cease to exist precisely when it acquires an ability. "That something should perish by virtue of *gaining* that ability [the ability to think] is absurd" (1997, 80). My argument in the text focuses on the counterintuitive consequence that an essential property of

The chief problem with this response is that it entails two ideas that pull it apart as a coherent response. The first is that an abortion *cannot* end the life of a person. The second is that the thing that can be killed by an abortion has the powers to become a person *had* the abortion not taken place. For we can ask: What kind of thing is it whose life is ended by an abortion? In raising this question, I'm suggesting that the self response must grant that abortion ends the life of what would *otherwise* be a person. What is this something? What are its qualities?

For convenience, let's grant that this something is a human organism, it has organs that function in characteristically human ways. It also has the feature of being able to either develop into a person whereby the person becomes a part of the organism (McMahan, 2002, 92 ff.) or the organism comes to constitute a person like a lump of clay may come to constitute the statue (Baker, 2000). Either option is entailed by the response because the response must grant that the person will exist if the abortion is not performed (assuming the organism does not die from other causes, e.g., pregnancy complications).

But if an abortion is the only thing that stands in the way of the organism coming to house or constitute a person, that tells us something about the nature of the organism. It tells us that on its own, it would develop and become or constitute a person. The fact that the organism has these causal powers is nothing short of amazing.

The logic of this response to the modal argument grants that what is killed by an abortion has the following features: (i) it has causal powers to become or constitute a person, (ii) it can do so in a continuous, self-directed pattern. Assume that a thing's powers are indicative of the thing's nature. One explanation for why it has such causal powers requires revising our starting hypothesis, namely, the organism does not become or constitute a person, but it might already be one. What might this explanation look like?

Devin Henry (2005) asks us to imagine a paper cup lying on the side of the road. After a few months that cup will have broken down into its constituent parts. It will be, after a while, merely pieces of paper and no longer a cup. But then imagine that these pieces of paper suddenly organize themselves into another paper cup, or even a lampshade. Suppose further that this amazing feat happens with surprising regularity. Paper particles organize themselves into paper cups. Henry's thought experiment is meant to direct our attention to how amazing self-organization is. It is precisely this property of self-organization that is evinced by embryogenesis. Not only are early-stage organisms self-organizing, but they also hit their mark—the embryo seems to "know exactly what it wants to be" (Henry, 2005, 8). So, the embryo organizes itself, not in any direction whatsoever, but in a very controlled and to a precise endpoint, namely, the mature form of its nature.

Mechanistic views will account for the development in terms of the biochemical matter causing ever higher levels of organization, complexity, and powers.[14] But for

---

a person is the complement of *being the victim of an abortion*. The text explains why this should be viewed as counter-intuitive.

[14] For why Aristotle's teleological views are compatible with modern biology, see Austriaco (2004) and Austriaco et al. (2024).

Aristotle, order at any lower level of material order cannot cause ever higher levels of organization. Jonathan Lear remarks that, "What is needed in addition is form as a basic irreducible force—a developmental power, …" (Lear, 1988, 39). The substantial form is not merely the mature adult form of an organism, it is certainly not another part such as a homunculus, and it is not reducible to genes as on genetic determinism; but it is that which accounts for why the organism is what it is and develops as it does. Again, Lear is on point,

> [I]t is a force in the organism for attaining ever higher levels of organization ….the idea that the order which exists at the level of flesh would be sufficient to generate the order required for human life was as absurd for Aristotle as the idea that the order that exists in a pile of wood would be sufficient for the pile to turn itself into a bed (Lear, 1988, 39–40).

In explaining this view, Aristotle often exploits analogies with artifacts and craftsmen. The analogies are apt in one sense; they tell us what it is to come into being and in virtue of what. The difference between artifacts and natural objects is that natural objects have their cause internalized. Nature is an inner principle of change and rest, making the thing be what it is. A thing's form is what the entity is. It follows that a natural object has its form, making it be the kind of thing it is (Lear, 1988, 17).

If you and I are individual entities with a nature/form, you and I exist when that entity does. Development does not introduce new entities; when you develop arms, legs, fingerprints and psychological powers, other things do not come into existence. Reaching a developmental milestone does not change the nature of that which reaches it. Mathew Lu states the point as follows,

> The relevant issue here is that any (putative) potential must belong to a substance with a particular nature. To say that a particular substance has a potential to develop in some way is not to make a prediction about the *future*, but to make a claim about that thing's nature *right now* (Lu, 2018).

As a human being develops through the embryonic, fetal, neonatal, etc., phases, *it* develops rational abilities. To say otherwise is arbitrary (see Sect. 4.5.2).

The plausible intuition is that development does not change the identity of the thing that develops. The reason why is because the change is internal in the sense that the human being causes its own changes with a view towards survival, growth, and maturation of its prototypical powers. A tree does not develop into a bookshelf; but a sprouted *platanus orientalis* seedling develops into a mature sycamore tree. A Polaroid picture of a jaguar develops from a brown amorphous smudge into a visibly clear picture of the jaguar that it was all along (Stith, 2014). Blank film does not develop into anything.

The idea that development does not change the identity of what develops does not suggest backward causation in which a mature human causes the immature human to develop. Rather it is to suppose that purposefulness exists in things. A mature organism of its kind manifests more species-typical powers than its immature counter-part; but those powers/capacities *express* what kind of organism the thing is, but they do not *cause or make* that organism be what it is.[15] Those powers are ontologically dependent upon the kind of substance it is (already), not vice versa. Consider the maturation from infant to adult. Significant cognitive and bodily changes take place in one and the same thing. As such, the end towards which x develops, tells us what x is all along.[16]

Though brief, it should be clear that the Aristotelian explanation offered here is *possible*. On the self response, comp-V is an essential property of persons. So, if it is possible for persons to possess V, the self response is undercut. The Aristotelian explanation supposes that being a person (already) is what explains why the organism has the causal powers it does. The person can exist before the full maturation of those powers. If it is possible that you and I are persons in this sense, then you and I would exist when an abortion could take place. It follows that it is possible that you and I could be victims of an abortion. Therefore comp-V is not an essential property of persons. The self response grants that the organism that could (later) be the victim of an abortion has the powers (i) and (ii). But a plausible explanation for why it has those powers is that it is 'already' a person—understood as a rational substance.

This rejoinder to the self response relies on two claims: the Aristotelian explanation for why the organism has the powers it does is because it is a person already. And it is possible that persons (you and I) are entities or organisms with a rational nature; you and I are not best identified with a specific power—even if that power is consciousness. Their conjunction is certainly possible. It follows that you and I *can* be victims of an abortion.

The rejoinder to the self response offered here is a species of a more general complaint. Whereas it is true that persons cannot be or even become chickens; when the same organism develops along a determined trajectory, it becomes philosophically harder to divide up the organism into different substances or entities even if what develops are psychological abilities. A general problem with any dualism enters at just this point: such dualisms must argue that unified conceptions of the human person (e.g., animalism or hylomorphism)[17] are necessarily false. One may disagree with these theories or not understand them, but to argue that they are necessarily false is a tall order. The next sub-section considers at some length an argument for why this is a tall order, namely, the too-many thinkers problem. Before addressing that problem, some minor observations are worth noting.

The hylomorphic view countenances the importance of having psychological capacities. But having psychological capacities counts as *evidence* for what kind of thing this organism is. Such capacities do not make a thing be what it is. Guilt is

---

[15] Cf. Baker's understanding addressed in Sect. 4.4.1, fn 15 to the effect that the capacities of a thing *makes* it be the kind of thing it is.

[16] See Napier (2015, 670 ff.) for a reply to McMahan's dilemma (2002, 11 ff.) Concerning substance accounts of the person, see Henry (2008). See Bertalanffy (1949) for why the account I describe here is neither preformationist nor epiphenomenalist. See Tollefsen (2010), Beckwith (2011) and Moller (2011) for further commentary.

[17] I do not consider hylomorphism a dualism. See Scaltsas (1994) and Austin and Marmodoro (2017).

having performed an illegal action, but evidence for guilt is, for example, finger-prints on a gun, nefarious motives, and so on. Likewise, manifesting psychological capacities is evidence for being a person, but those capacities are not what a person is. Viewing them as evidence is quite plausible, and it retains the intuition that such capacities are ontologically important.

A second observation is that the self response does not grant the intuition that Sarah can truthfully assert P″; it simply denies it. Here, the proponent of the self-response must argue that if there is a defect, there must be equally plausible justifi-cations for thinking that Sarah is confused (viz., a justification-defect satisfying condition (ii)). But those supposedly 'equally' plausible justifications are the psy-chological accounts which I have argued are underdetermined; they are not equally plausible. Given the previous results (i.e., Chap. 4), the modal argument does not suffer this defect.

More can be said in this regard, and the next section takes up what is known as the too-many thinkers problem for any such body-self dualism.

5.4.3   The Too Many Thinkers Problem

The self response asks us to suppose that you and I are different things from our bodies. Some argue for this distinction because self and body have different persis-tence conditions (Shoemaker, 1999; 2016). But as was previously noted, claiming that those two things have different persistence conditions requires identifying those two things as distinct already. And so, the distinction between self and body is pre-supposed, not argued to. In any case, set aside the reasons for the distinction and just consider that on such a view you are not the same thing as your body. You are, instead, a part within the broader whole of your physical organism (McMahan, 2002), or you are coincident with your physical organism (Baker, 2000).

On either account we can ask what happens when you think a thought? Does your human organism also think that same thought? And if you and your organism are two different things, then two different things are thinking the same thought. It is not too radical to suppose that your organism thinks the same thoughts as you, after all, your organism has the same brain as you, the same nervous system as you, and so on. As Olson explains,

> Yet the animal would appear to be rational and intelligent. At any rate it has a normally functioning adult nervous system. It is physically indistinguishable from you. It has the same surroundings and history. What more does it take for a thing to be able to think?″ (Olson, 2002, 190).

If we are non-reductive materialists with regard to the mind, it follows that when you think a thought your human organism thinks that thought as well. We now have one too many thinkers than there should be. This is the too many thinkers problem.

There is an additional problem for the self response, namely, you could not *know* that you are the self versus the body. Again, Olson explains what he refers to as the epistemic problem as follows (2002, 190).

> You may think you're the person. But whatever you think, the animal thinks too. So the animal would seem to believe that *it* is a person, with psychological identity conditions. It has all the same reasons for thinking so as you have…. If you *were* the animal and not the person, you'd still think you were the person.

Externalism with respect to knowledge, according to which knowledge requires having reliable cognitive faculties, does not obviate this problem because, crucially, one would need to identify *whose* faculties in question are producing the belief. The same ambiguity that Olson is referring to recapitulates one level up, so to speak, to whose faculties are generating the beliefs. When I form the belief that I am a person, is it the animal's faculty or the person's? Again, insofar as the animal and the person share the same brain and putatively, the same processing mechanisms, we can ask: to whom do these cognitive faculties belong?[18] You, whichever one you are, cannot know because your evidence base between the two hypotheses (I'm a person vs. I am the organism) are the same.

So there are two problems: the too many thinkers problem, and the epistemic problem. These problems function here as justification defects (condition (ii), "there are equally plausible justifications for ~P"). The intuitive evidence highlighted by these problems function as countervailing or counterbalancing evidence against a psychological account of the person. Of course, Olson intends that these problems function as arguments for animalism, but I intend them here to function merely as justification-defects against the self response.

One response to the too many thinkers problem claims that the relationship between person and human organism should be understood as part to whole. McMahan (2002) gives several examples to illustrate. When a limb on a tree grows, the whole tree is said to grow. The limb (the part) grows, and (the whole) tree gets bigger *in virtue of* the limbs growing. It is not absurd to say that there are two things that are growing, the limb and the tree. The tree is not constituted by the branch (cut the branch and the tree still exists), but the branch is certainly *part* of the tree. Likewise, the horn in my car makes a noise, but the car makes a noise *in virtue of* the horn making a noise. "In the same sense in which the tree grows because its limb does, and in which the car honks because its horn does, my organism may be said to think, feel, and perceive because I do" (McMahan, 2002, 92). And he states that these "analogies help elucidate the sense in which there are two conscious entities present where I am. My organism is conscious… only by virtue of having a conscious part" (McMahan, 2002, 93).[19]

---

[18] Olson responds to a similar objection, but his response assumes an internalist view of justification. As is clear in my response, one does not need to assume internalism. It is sufficient to point out that one could not know, on body-self dualism, whether externalist conditions are ever satisfied.

[19] McMahan makes clear that his solution to the too many thinkers problem is an improvement upon Baker's solution according to which the animal thinks derivatively, but the person thinks non-derivatively. A corollary is that the animal runs or walks non-derivatively, and the person performs

The first observation to make is that understanding the relationship between the self and body as part to whole does nothing to assuage the epistemic problem. In virtue of what, one might ask, can the part say that it is the person? In virtue of what evidence might the person not be justified in saying that *he* walks, or digests food, or stubs *his* toe? Rather, he has every reason to believe that his toe is stubbed, suggesting that he is the whole and is not located in a part.

Furthermore, the limb and the tree analogies are not helpful for understanding McMahan's view because consciousness is not located in a particular part of the human organism. A functional brain causes one to be conscious, but McMahan would be the first to resist the idea that *I am located in my brain* as a quasi-homunculus. He criticizes other views for implying that persons are homunculi (2002, 14ff.). As Bennett and Hacker observe,

> thinking, believing, deciding and wanting, for example, cannot be assigned a somatic location. The answer to the questions 'Where did you think of that?', 'Where did he acquire that strange belief?', 'Where did she decide to get married?', is never 'In the prefrontal lobes, of course' (Bennett & Hacker, 2008, 250).

Likewise, the idea that *I* am located somewhere in this body is equally nonsensical. The activity in my brain might be a causal condition for *me* being conscious, but not that *my brain* is conscious. Again, Bennett and Hacker bring clarity: "It is human beings…, not their brains…, that fall asleep and later awaken, that are knocked unconscious and later regain consciousness" (Bennett & Hacker, 2008, 246). It is certainly true that neuroscientists are busy identifying the areas of the brain where various cognitive operations occur. But "they are discovering which parts of the cortex are causally implicated in *a human being's* thinking, recollecting, deciding" (Bennett & Hacker, 2008, 251, emphasis added). To say otherwise is nonsensical for the brain does not think or decide, I do.

McMahan's 'in virtue of relation' functions in the same way that Baker's distinction between derivative and non-derivative functions in her response to the too many thinkers problem. Both distinctions are meant to explain how it is that only one thing thinks even though that one thing is part of or constituted by another thing, the human organism. So, my organism thinks but only in virtue of me thinking. Its status as a thinker, is causally and ontologically parasitic on me thinking. Would we say the same thing, however, about corporeal properties? Should we say that "I hurt in virtue of my foot hurting"? That would be an odd view since when I stub *my* toe, *I hurt*.

The part-whole distinction is oblique in other ways as well. The suggestion is that I am a part of my organism. Which part? Clearly, consciousness is not a part of my organism, at least not like my foot is a part of my organism. A candidate for which part I might be on this view is a functional brain. But if I am a part of my

---

such activities derivatively. It strikes me, however, that McMahan's 'in virtue of' relation is similar enough to the derivative relation such that the criticisms mentioned in the text apply to both approaches. And even if they are not, Baker's view is scrutinized elsewhere by Eberl (2006, 2020), Burke (1994), and Wiggins (2001).

human organism and that part is a functional brain, then I am a functional brain. That view would lead to absurdities. For one, I am 3.5 pounds, not 180 (Olson, 2007). My wife can only kiss me during neurosurgery (Pruss, 2011). I can only be injured by way of a concussion or, again, during neurosurgery. I cannot be raped or be a victim of battery as those crimes are typically understood. I don't ever run or walk though I might accompany my organism when it runs or walks. If I were my brain, I do not *perform* actions at all. (I might only intend them but even that seems odd since my brain does not intend, or survey options, I do).

Consequently, we need greater clarity on several aspects of this view, but I'll focus on what the 'in virtue of' relation might mean. As I see it, it can be understood in two different ways. Consider the following two propositions:

(Causal): The airplane moves *in virtue of* the engines.
(Mereological): The engines fly *in virtue of* being a part of the airplane.

On the causal understanding, the airplane moves in-virtue-of the engines; i.e., the engines cause the airplane to move. On the mereological understanding, the engines fly only because they are attached to the airplane. Without the empennage and wings, the engines cannot go airborne.

Applied to persons and their organisms we can say the following. On a causal understanding of the in-virtue-of relation (which is suggested by the car horn analogy), my organism is caused to think by me every time I think. This suggests that my organism has psychological states and properties every time I think. It is agreed, however, that my organism is not the same thing as me on this version of body-self dualism. Since I have psychological states and properties, it follows that there are two different things that have the same psychological states and properties (my part, and the whole of which I am part). When I cause my organism to think, it thinks. But what does it think? Presumably, it thinks the very same things I do. But if my organism and I are two different things, McMahan still must grant that there are two thinkers.

It is true for an artefact that when the horn sounds, the horn is making the noise, but it is a part of the car. Saying that the car honked is just a loose way of saying that its horn sounded. Maybe the thought applied to persons and their bodies is this: it is odd to say that Alice's mind weighs 3.5 pounds (the weight of her brain), or that Alice's body is reading (Olson, 2003, 328). Alice has a mind and a body, and the two are different.

In response, these observations are metaphysically innocuous. It is not odd to say that Alice thinks and that she walks. As pointed out in the discussion of Hacker (2009) in the previous chapter, Alice has psychological properties, and corporeal ones too. When she thinks, *she* thinks; and when she runs, *she* runs. There is no reason to reify her mind into a separate entity we call person or self. The mind is a useful *facon de parler* for a human being's psychological capacities.

Suppose that the in-virtue-of relation is understood as mereological (which is suggested by the limb analogy). My organism thinks only because I (the thinker) am a part of the organism. Does my organism have any psychological properties if we imagine it without the thinking part? Either it does or it does not. If it does, then what psychological properties does it have and why be so sure that it cannot be a person? If it does have psychological properties, where are the corresponding psychological powers? David Armstrong observes that all properties bestow a power (active or passive). He says "it is only in so far as properties bestow powers that they can be detected by the sensory apparatus or other mental faculty" (1980, 45). We can ask, then, whether the organism has the property of being a thinker. If it does, then it would have the power to think. On the supposition that organism O and person P are different things, there would be two thinkers. Hence, the too-many thinkers problem remains.

If it does not have psychological properties, i.e., being a thinker, then it would not have the power to think. On this latter supposition the organism would not have any psychological states, properties, or powers, and this is to court radical dualism of which McMahan rightly rejects. When the whole of which I am a part has different essential properties, it is a different thing. What generates the problem for the mereological interpretation is that being a thinker or being conscious is something a thing either possesses or it does not.

So, if the 'in virtue of' relation is understood as causal, McMahan has not obviated the too many thinkers problem. If the in-virtue-of relation is understood as mereological, he is implicitly committed to radical dualism, which he explicitly rejects. Another way to motivate the latter claim is to consider the nature of the part in the part-whole relation. McMahan wants to understand that persons or selves are parts within the whole, i.e., the human organism. But is this part merely a psychological capacity, or is it a separate entity? If it is a separate entity, radical dualism follows. If it is merely a psychological capacity, it is counterintuitive to say that I am identical to it—as has been repeated numerous times in this discourse.

One might think of the general point this way. We say of an entity that it can move, develop, or think. But it makes no sense to say that consciousness moves, develops, or that it thinks. My consciousness does not think, I think. I might say that my consciousness develops in some religious or mystical sense, but that is not what psychological theorists mean. The point is that consciousness is not a thing or an entity; it is a power of a thing, akin to intellect and will. I can become conscious of something, suggesting awareness. I am conscious right now, as opposed to being unconscious, suggesting that I am in a state of consciousness. I can lose consciousness, as when I fall asleep or get hit. But I continue existing. Of note, when I get hit, my consciousness does not get hit, I do.

So there are numerous cases in which 'consciousness' and 'entity' or the first person pronoun 'I' are *not* interchangeable *salve veritate*. The linguistic evidence strongly rules against reifying consciousness into a thing or identifying that power with me. Persons might have consciousness (an ability) or be conscious (a state of awareness), but consciousness is not a thing, it cannot *be* me.

A similar confusion pertains to identifying persons with their functional brains—as has been discussed. The point to raise here is that such a position involves a mereological confusion. Again, Bennett and Hacker brings clarity on this issue.

> Human beings have brains, but human brains do not have brains—they *are* brains. Human beings have bodies, but a human brain does not have a body, it is a part of a human body. Human beings have minds. But brains don't have minds (there is no such thing as the brain making up its mind, having a dirty mind, or having a thought cross its mind). The fact that human beings have the distinctive capacities constitutive of having a mind is dependent upon the normal functioning of their brain. That no more shows the brain to be a human being, or a limiting case of a mutilated human being, than the fact that an aeroplane's capacity for flight depends upon the normal functioning of its engines shows that the engines fly, let alone that an engine is an aeroplane or a limiting case of a damaged aeroplane (Bennett & Hacker, 2008, 243).

My consciousness does not think, remember, decide, or hope. Rather I do. My brain might be in a sclerotic state but I can't be sclerotic. I may be cheerful, but my brain is not cheerful (Hacker, 2009, 252). I can have properties that neither my brain nor my consciousness has. The inference of course is not that I am some mysterious separate thing or a bare particular. The conclusion that follows is that the thing that thinks, remembers, and so on is *this* human being. That same thing can be cheerful, groggy, hungry, light on his feet, and so on. There is one thing, the human being, and this human being has both corporeal and psychological properties. I, however, am identical to this human being. No absurdities follow by supposing that I am this spatiotemporal continuant; this human being.

Even if clarity can be given the relationship between person and human organism on some version of a body-self dualism, and supposing that this clarity can be understood via a part-to-whole relation, *we have no reason for thinking that the part does not come into existence when the whole comes into existence*. As is also repeated numerous times in this discourse, the part understood as an exercisable capacity for consciousness is one among several ways of understanding this part of the whole. (If it is a part, it does not matter whether it is exercisable or not qua part). The part that grounds this capacity for consciousness can be traced back to the zygote having the requisite DNA that codes for its development. Again, nothing prevents us from saying that even if you and I were parts of our organisms, that part comes into being *when* the whole does. How so? We may understand part concretely as in this functional brain—in which case the absurdities previously noted result. Or we may understand it more abstractly as a capacity for consciousness which may exist as early as immediately after successful conception. The point is that understanding you as a part within a whole so as to obviate the too-many thinkers problem also tolerates understanding you as coming into being at or shortly after conception.

We could equally understand the relevant part as a capacity the whole has because of the kind of whole it is—because of the nature of the thing it is. On this latter understanding, when the whole thinks, it thinks. When the whole develops an exercisable capacity for consciousness, the whole is conscious. There is not another thing that comes into existence when the whole develops *its* prototypical capacities. On this view there is one person who possesses both psychological and corporeal properties and whose prototypical powers develop over time. But it is the person who develops those powers rather than the person coming into existence only when those powers manifest. Obviously, on this option, a body-self dualism does not follow and the modal argument is confirmed.

5.4.4   The Modal Argument and Prenatal Harm Cases

Consider the impairment intuition from Chap. 3 again. There we discovered that when an agent impairs a human being prior to the onset of consciousness, we still have the intuition that the human being was harmed. Furthermore, when the harm occurs is when the injury occurred. It would be odd, for example, to say of Persson's neuron removal case (Sect. 3.3) that the preconscious human being was not harmed until he *experienced* the devastating effects of the pernicious brain surgery. That case involved fetal neuro-surgery—supposing that this is compatible with the survival of the child. Suppose the surgeon removes neurons causing the child to be mentally disabled. Certainly, in the cases in which the impairment occurs and the injured survives, it is natural enough to say that the child was harmed *by* the neuron removal. If I did not exist prior to the development of consciousness, I can't be harmed either. But the impairment cases reveal, especially in cases of survival, that it is counterintuitive not to suppose that I exist prior to the actual exercise of consciousness. For otherwise, the intuition that impairment involves a harm does not stand.

The proponent of the self response may argue that there is nothing counterintuitive about the injury and wrongdoing being temporally disparate. Consider an evil surgeon who genetically modifies the gametes of someone so that the gametes are more susceptible to disease if conception were to occur. It might be the case that such modifications are wrong for any number of reasons, but most will agree that it is not *the gamete* that is wronged. If conception were to occur with those modified gametes, the (future) person would be wronged but only when that person comes into existence. Here is a case where the injury and the wrongdoing are disparate.

Though the case illustrates how injury and wrongdoing can come apart, the explanation for the *wrongdoing*, however, must borrow assumptions ensconced in a hylomorphic view of the person, specifically the idea that development does not alter the identity of what develops. What links the injury with the wrongdoing in the gamete-modification case is that we are presupposing certain facts about conception *cum* continuous human development of that entity. If different ontological entities crop up, especially purely mental ones, there's no reason for thinking that manipulation of the gamete (a biological entity) would wrong a person understood as a psychological entity. There is no reason to think that there is a link between the nature of the injury and wrongdoing of the *person* if one did not presuppose such facts. So, though the hylomorphic view does not hold that you and I are identical to gametes, it does hold that *conception* is the event that links the injury with the wronging. In fact, it *must* suppose this. Had conception not occurred, there is no thing that is harmed in the gamete-modification case. If it does occur, there is something that is harmed, (assuming there are no pregnancy complications). Because there is post-conception a new entity that develops in a continuous and self-directed manner, conception must be understood as effecting an ontological change sufficient to render the gamete-modification *as* wronging someone.

5.5   Conclusion

Let's summarize the dialectic to this point. One view of the person, the functional view, requires that persons must be able to exercise psychological potencies, in particular consciousness. Because the hylomorphic view of the person also countenances the importance of rationality, and having such potencies, it too can grant the intuitions revealed by the brain-transplant cases, the body-transplant case, and the dicephalic twin example. As phrased in the previous chapter, the difference between the two approaches is small enough to countenance the same intuitions, but big enough to rule differently on *when* the person comes into being. For the hylomorphic view, persons are entities that have psychological powers, but these powers need not be immediately exercisable, but possessed because of the kind of being he or she is. That personal identity could track biological development is illustrated by the brain regeneration cases (Sect. 4.4.1). And, similarly, that persons can come into existence prior to a specific episode of consciousness is illustrated by the embryo-brain transplant cases (Sect. 4.4.2).

This chapter presents an argument based on a thought experiment the intuitions of which my interlocutors should share as the experiment simply extends the basic idea informing prenatal harm cases. Putative de jure defects of the modal argument are twofold: it begs the question and it suffers equally plausible counter-considerations (i.e., the self-response). I capitulate to the first charge, but with the qualifier that the circularity charge rests on an unjustified burden of proof, one that my interlocutors would not apply to either their own arguments, or to arguments against infanticide or toddlercide.

The self-response is rebutted because it is too strong, and the response itself is rebutted by the too many thinkers problem. Explicit reflection on prenatal harm cases against reveals that the modal argument is merely a limit case of them.

The conclusion that follows is that the arguments for (pk) suffer from serious de jure defects of the inference variety (vicious circularity, the problem of underdetermination, or under motivation). The modal argument for ~(pk) suffers from a justification defect but the argument that it does so suffer from the defect itself suffers from several defects of its own (5.4.2–5.4.3). The arguments for ~(pk) suffer from much weaker de jure defects compared to the arguments for (pk).

In the next chapter we turn to analyze (pk) as it is understood in the context of euthanasia. Since issues of personhood rarely enter the euthanasia debate, (pk) will represent the position that it is permissible to kill one's patients who are, on most criteria, persons.

Chapter 6
Euthanasia, Suffering, and the Intention to Kill

6.1 Introduction: Suffering and the Argument for Killing

The previous three chapters discussed various arguments for and against the permissibility of killing unborn human beings. Over the next two chapters we discuss the permissibility of killing adult persons who have voluntarily requested killing. (Henceforth, (pk) is understood in these two chapters as referring specifically to euthanasia.) The present chapter canvases the arguments for (pk).

For the abortion debates ontological questions about the status of embryos and fetuses loom large, but for euthanasia, the personhood of what is killed is typically taken for granted. On most conceptions of what is required for personhood, even the demented or minimally conscious patient counts as a person.

What is at issue in the euthanasia debate is whether the moral prohibition against killing innocent persons admits exceptions *in cases where such persons request to be killed because death would be better than continued living*. Different authors use different phrases following the "because" clause such as, death would benefit the person, or the person would be better off dead, or the person does not and would not have a life worth living. These are all cognates of the same idea that continued living would be worthless, or the reasonably expected burdens of continued living outweigh the benefits and therefore life is not worth preserving. This is the chief axiological claim serving to justify an exception to the general prohibition against killing innocent persons.

The argument for (pk) usually centers around two values, namely, the value of self-determination and the value of avoiding a life that is not worth living. Avoiding a life that is not worth living can be understood simply as wanting to avoid present or future suffering. Some authors use the term well-being to refer to this value (Brock, 1992). Such nomenclature is confusing since suffering is contrary to one's well-being, as is not existing in so far as it eliminates well-being's very possibility, viz., there is no *being* who can be *well* when one does not exist. Though the nomenclature is confusing, what it means is shared across most authors, namely, the value

S. Napier, *Justified Killing*, https://doi.org/10.1007/978-3-032-14946-6_6

of preserving one's well-being applies to one's choice to avoid a life that is of low quality, or is unacceptably burdensome, or is fraught with intractable suffering.[1] A life with such experiences is not worth sustaining. This is enough to understand the principal argument for (pk) as follows:

1. It is permissible to kill an innocent person S if and only if (i) S requests to be killed and (ii) S is suffering harm to the degree and kind that S judges unbearable. "When harnessed together, the value to individuals of making autonomous choices, and the value to those individuals who make such choices of promoting their own well-being, provide the moral foundation for requests for voluntary euthanasia" (Young, 2022, §3).

Of course, the term 'harm' must be understood in a specific sense. One might think that killing just is an instance of harming. So, how could suffering harm justify killing if the latter is a case of harming itself? Understood as such the first premise says that it is permissible to harm someone if and only if the person is suffering harm. That can't be right. It looks like this premise must beg the question from the start. It must *assume* that killing is not a case of harming. If killing is wrong, it cannot be so because it harms on this assumption. If killing is not a case of harming and the one killed wants to be killed, there appears no other reason for opposing it. Killing someone is neutral on this view and is only made wrong if it conflicts with someone's will. If one already assumes this line of thought, the argument is obviously circular. We might patch up this defect by replacing 'harm' with something like; 'suffering a setback in one's interests that one judges unbearable'. But this patchwork does not get us far. If the term interests are understood subjectively, then (1) entails the permissibility of suicide. I flag these problems only to motivate why some deferred maintenance is required, as is done below.

2. Euthanasia is the act of killing people who satisfy conditions (i) and (ii).
3. Therefore, euthanasia is permissible.

The defense of premise (1) has an extended genealogy behind it. Its defense begins with mid-twentieth century developments in value theory. C. I. Lewis, discussing intrinsic value, says the following, "In this sense of 'intrinsic value' as the value of that which is valued for its own sake, no objective existent has strictly intrinsic value; all values in objects are extrinsic only" (Lewis, 1971, p. 387). By "all values in objects are extrinsic only" Lewis means that intrinsic value is located if you will, only in the goodness of experience.

> The goodness of good objects consists in the possibility of their leading to some realization of directly experienced goodness. What could by no possibility ever be an instrument for bringing any satisfaction to anybody, is absolutely without value, or the value of it is negative. Hardly anyone would deny this (Lewis, 1971, p. 387).

---

[1] Stating the argument in these terms follows Young (2022) who is discussed in the text. The empirical reality is quite different. The most frequent cited reason for requesting euthanasia is not pain, but loss of autonomy (91.5%). See "Oregon death with Dignity Act – 2014" DWD2014ARnarrative20150202x (accessed 10 January 2025).

Playing the same theme, Brand Blanshard argues that consciousness alone is intrinsically valuable. "Moore's final conclusion was, then, that a universe without consciousness would be a universe without value. With this I think we must agree" (Blanshard, 1961, p. 273). Further on, Blanshard offers the following riposte to the notion that things have value:

> Thomists, for example, continue to hold the obscure view of Aristotle that each material thing is a more or less complete actualization of what it is potentially, a more or less perfect embodiment of its special form, and is, so far, good. The disappearance of a boulder on some unheard of dark star would therefore be a loss of value to the world. Of course one may conceive of value in this way if one wishes, but it carries one quite out of touch with ordinary meaning; most men would say it made no difference whatever to the amount of good in the world whether such a boulder existed or not. Take any example of what we ordinarily regard as good or bad, imagine consciousness away, and the values vanish with it (1961, 273).

Though a brief survey, two key claims emerge from mid-twentieth century value theory. The first is that *things* do not or even cannot have intrinsic value—though this is challenged by later developments in environmental ethics. Second, the only subject of intrinsic value is a conscious state, specifically, one whose content is a good experience. Slight permutations or specifications of this claim emerge in contemporary value theory under the phrases "satisfaction of one's preferences," "fulfillment of one's desires," or "realization of one's time relative interests." Though these phrases do not mean the exact same thing, they are, I'm suggesting, members of the same family insofar as having pleasurable or fulfilling (i.e., satisfaction of one's interests) experiences is intrinsically valuable.

James Rachels serves as a representative vinculum connecting the axiological commitments voiced by Lewis, Moore, and Blanchard, to the issue of euthanasia specifically. He states that "from the point of view of the living individual, there is nothing important about being alive except that it enables one to have a life [i.e., have a life in the narrative or biographical sense]. In the absence of a conscious life, it is of no consequence to the subject himself whether he lives or dies" (Rachels, 1986, 26). One can see in this passage, explicit mining of Blanshard's insight according to which "imagine consciousness away, and the values vanish with it."

Rachels makes two 'moves' in this passage. The first is to disaggregate two notions of a life. The first notion is a mere biological notion according to which someone is said to be alive. What is living, on this first notion, is a biological organism. Other authors offer us various flats and sharps on the same note. McMahan has us suppose that "each living individual is essentially a member of its own species, and that in that sense an individual's species membership indicates its essential nature. Even if that is true, there is no reason to suppose that its species membership determines its moral status" (McMahan, 2002, 216). Marvin Kohl writes that the value of life on this first account is the worth that attaches to a human being "just because he is a member of the uniquely rational and capable species" (Kohl, 1975, 133). Kohl, anticipating McMahan's assessment, argues that species membership is of no value. Shelly Kagan (2018) helpfully refers to this view of life's worth as the "neutral container view." Biological life itself merely enables me to have a good (or bad) life (life understood to include one's experiences, plans, and projects).

The second notion of a life is a biographical notion according to which someone has a life or lives a life. On this second notion, having a life means having a story to tell, it means having a narrative which includes the fulfillment of one's plans and projects. In contrast to the first notion, we intuitively resonate with the value of a good biographical life. Rachels' point is that the prohibition against killing should only apply to the killing of someone with a good biographical life. "A person in irreversible coma, or an infant with such defects that it will never mature, is not the subject of a life [in the second sense]; so they fall outside the scope of the rule thus understood" (Rachels, 1986, 28).

The second 'move' in the quotation above is to give an argument for non-voluntary, active euthanasia. Such a position, however, should not be viewed as heretical if one accepts that death can be a benefit. If death is a benefit to patients in certain circumstances, why wait for the patient to ask for it? Healthcare practitioners benefit their patients in numerous situations without the patient asking for that benefit, e.g., most patients who are cared for in the ICU. So, if death is a benefit, requiring that the patient ask for it deviates from the usual practice of healthcare practitioners benefiting their patients without explicit consent (see Keown, 2018).

The defense of voluntary euthanasia simply adds the requirement that the agent must judge that his or her life is not worth living and requests euthanasia to end it. If consciousness is the only value and if it is filled with only bad experiences, ending it does not end something of any worth. Rachels provides the needed vinculum here. "If a person prefers… death as an alternative to lingering on *in this kind of torment*, only to die anyway after a while, then surely it is not immoral to help this person die sooner" (Rachels, 1986, 154). The torment to which Rachels refers is the torment of going through a dying process that is particularly painful and fraught with unmitigated suffering. I suspect Rachels envisions that at least some cases of dying will exceed the abilities of modern palliative medicine. That assumption explains why he understands the issue as follows: "The moral issue is whether mercy killing is permissible *if* it is the only alternative to this kind of torment" (Rachels, 1986, 154).

I must observe here that Rachels' presentation is a bit disjointed. In the quotation above, he refers to patients who are in an irreversible coma, they are unconscious. If they are unconscious they cannot experience "torment" because they cannot experience anything in so far as they are unconscious. So, if it is the *experience* of suffering that supposed to justify killing, the alternative is palliative sedation. Killing is not the only alternative. If it is the absence of consciousness that justifies killing, then why focus so much attention on the experience of suffering? Maybe there are two features of patients that would justify killing them: the absence of a *meaningful* life because of permanent unconsciousness, or the presence of intractable suffering understood as *experiencing* psychic or physical pain. I set these ambiguities aside for the moment.

Rachels' discourse constitutes a tradition of sorts that is picked up by later defenders of (pk). Robert Young remarks that "a person can be better off dead and believe this to be true when the life that remains in prospect for that person has no positive value for her" (Young, 2022, §1). Daniel Brock remarks that,

> If self-determination is a fundamental value, then the great variability among people on this question [the question of whether the value of self-determination extends to choosing one's death] makes it especially important that individuals control the manner, circumstances, and timing of their dying and death (Brock, 1992, 11).

When discussing the value of well-being, which initially appears as a value pertaining to the quality of one's life, the value of self-determination is given voice again. The judgment that life itself can become a burden and thereby death a benefit, is a judgment that,

> underlies a request for euthanasia: continued life is seen *by the patient* as no longer a benefit, but now a burden. Especially in the often severely compromised and debilitated states of many critically ill or dying patients, there is no objective standard, but only the *competent patient's* judgment of whether continued life is no longer a benefit (Brock, 1992, 11, emphasis added).

The point of the distinction between life considered biologically, and life considered biographically is to argue that what is truly important is protecting someone's interests, their plans and projects. If someone does not have an interest in continued living the person can thereby waive his or her right not to be intentionally killed. If the reader does not like the language of rights, focus instead on goods. Suffering can render one's life as not worth living, a corollary of which is that death would benefit the patient. If this is correct, it is compatible with competency for a patient to judge one's life as not worth living and with that, death would be a benefit. Thus, the prohibition against killing innocent persons applies only to those whose biography is on the balance good. When the quality of one's life deteriorates to a point where the agent judges that her life is not worth continuing, killing the person is permissible.

But why think a thing like that? Far from ringing true, these claims ring hollow. Consider, for example, a love lost teenager who requests assistance in dying. Nothing in what has been said so far gives us any reason for *not* following through on the person's request, but that is deeply counterintuitive. Most euthanasia defenders wish to carve out circumstances in which such requests make sense from those that do not, and that the love lost teenager is in the latter category. But what condition or criterion does the love lost teenager not have, but the stage IV cancer patient has? We can easily imagine that either their physiological or psychic pain are intractable. Both seem to satisfy the criteria in premise 1.

Lachs' position on how to distinguish these cases is as follows. He says, "Many observers of no more than average sensitivity…" (2010, 562) and "[e]ven people of ordinary sensitivity understand that…" (2010, 565) there is a difference. No reason is offered for why we should prevent the lovelorn teenager's suicide and yet aid in the terminally ill patient's suicide. The distinction, he claims, is a basic moral intuition. We are asked to just agree that certain patients may reach a threshold in which it is permissible to intentionally kill them. People of average moral intelligence should just see this.

Lachs is not alone in making such an appeal to moral intuition. Larry Sumner considers a patient he names Anita whose disease progression is causing intense suffering that cannot be managed short of terminal sedation. (He also considers Bill together with Anita, but Bill's suffering is *anticipated*.) After noting that rights function to protect someone's choices or interests, waiving one's right not to be killed is compatible with either function. If rights protect choices, one choice an agent might have is to be killed if she views her life as not worth living. If rights protect interests, an interest one might have is not suffering at the end of life. On either option, the right not to be killed can be waived by Anita. Sumner concludes, "If, *as we are assuming, requesting euthanasia would be both beneficial for Anita and Bill* and an exercise of their autonomy, then it would be perverse to insist that the inalienability of their right to life prevents them from doing so" (Sumner, 2022, 87, emphasis added).[2] As emphasized, Sumner hopes that his readers will just see that patients who are suffering at some threshold may benefit from being dead.

Alan Buchanan also considers two cases both of which involve competent and informed terminally ill patients, one for whom pain relief is "unattainable or would so impair the patient's cognitive abilities…that he finds this option incompatible with his conception of personal dignity" (Buchanan, 1996, 24–25). Withdrawing other means of sustaining life would, Buchanan supposes, merely draw out the length of the dying process. Buchanan thinks that euthanasia is clearly permissible in this case. He comments that the "problem about intending death, then, is the apparent contradiction between allegiance to the prohibition against intentionally taking innocent human life, on the one hand, and the conviction that, *at least in cases such as these*, intentionally killing is permissible" (Buchanan, 1996, 25, emphasis added). Buchanan takes for granted that it is permissible to intend to kill patients in certain circumstances. The problem, as he understands it, is how to make coherent our basic moral intuitions; but one of those basic intuitions is precisely (pk). Certainly, Lachs and Buchanan hold (pk) as a basic belief. Sumner is a bit more careful. What these quotations show is how hard it is to shore up the argument for (pk), as is brought into relief in 6.2 below.

To summarize, the contemporary defense of euthanasia combines two necessary and jointly sufficient conditions that make euthanasia permissible—viz., these conditions jointly justify an exception to the general prohibition on the intentional killing of innocent persons. Those conditions are that the patient is suffering intractably and has competently judged his or her life as not worth living, or that one's death is a benefit.

6.2 De Jure Defects

These statements of the argument for (pk) require further defense as some of the claims, as I argue below, do not do the work of justifying euthanasia. The first difficult task is to justify the *inference* in premise (1). Premise 1 reads:

---

[2] Sumner cannot mean that "requesting euthanasia" is what would benefit Anita and Bill. He needs to say that *performing* euthanasia would benefit them.

(1) "It is permissible to kill an innocent person S if and only if (i) S requests to be
     killed and (ii) S is suffering harm [or some cognate] to the degree and kind that
     S judges unbearable."

We need a reason for thinking that if somebody requests euthanasia in a setting of
suffering (at some unspecified threshold of degree and kind), it is permissible to kill
that person. The previous section closed by noting that many authors take for
granted that it is permissible to kill innocent persons in certain circumstances. But
what is the argument for this claim?

To motivate why there is a need to provide an argument, consider some basic
conceptual issues such as the meaning of the phrase "death is a benefit" or its cognates, "life not worth living." Without a concrete meaning, it is hard to see how they
can figure in a justification for (pk).

Consider the meaning of death being a benefit. How may not existing count as a
benefit? To be benefited, one must exist. Likewise, having a life not worth living can
mean different things. We might say that a sports figure caught in a career long dop-
ing or cheating scandal has a life not worth living. The life was a lie, his accomplish-
ments were fictions, and the motives behind his actions reflective of pathological
ambition. It is by no means permissible to kill such persons. A prisoner of war has
a life not worth living, but it is by no means permissible to kill such a person. What
do such phrases mean such that their meaning entails or justifies thinking that it's
permissible to kill an innocent person?

One way to understand such phrases is that they are comparatively evaluating
two (or more) different states of affairs. In one state of affair (SOA), a patient continues to exist either in a permanently unconscious state, or with consciousness of
one's unbearable suffering. That state of affair is contrasted with one in which
everything else in the world remains the same except that that person is dead. Death
is a benefit if one thinks that the disvalue in the first SOA is more than in the second.
Not existing is viewed as better than existing *cum* suffering.

That way of understanding things however would still not explain how *the subject* is benefited in the SOA in which the subject doesn't exist. When we compare
the two states of affairs, one dyad does not include the subject. Therefore, we cannot
say that death is *good for* the patient. We are not holding constant the same subject
of value across both states of affairs for in one relata, the subject does not exist. At
best we can say that from the third person point of view or objectively speaking, the
second SOA (death) has less disvalue than the first (i.e., the *subject's* suffering);
conversely, the second is better than the first. This understanding does not have the
implication that *the subject* is benefited by being dead. But it does entail that certain
states of affairs can be valuable or disvaluable *apart from* a subject's interests,
desires, or conscious states—because in death, there are no such states. The argument for (pk) must accept a broader horizon of axiological commitments than was
initially advertised. In any case, these phrases should be understood as something
like a comparison of SOAs. I assume that understanding in what follows.

What follows here are extended descriptions of five de jure defects of the argument for (pk).

6.2.1   On the Distinction Between Biological Life v. Biographical Life

Before addressing any defects, we need to explore what this distinction might even mean. An initial statement of the problem is that the distinction is arbitrary in the sense that it is made in light of one's interest (Habermas, 1972). One is not "cutting nature at its joints" in the sense of making a distinction that corresponds to reality. The basic idea is explained by Alexander Pruss.

> It is also important to remember that to say that we are organisms is not to identify the metabolic functions as what is central to us. After all, we are not simply organisms. We are also, among other classifications, eukaryotes, chordates, tetrapods, amniotes, mammals, and primates. There is no reason to think that it is only what is found at the highest taxonomic level, that of organisms, that is central to our animal functioning. *It is also in our nature, as the kind of organism we are, to develop a functioning cerebrum which is a central control system. We are rational animals* (Pruss, 2011, 16).

One way to state the problem is that the distinction between biological and biographical life is that it is *artificially* narrow, it does not include all of what we are qua rational animal. As one ascends the classification Pruss notes, there is no reason to stop at mere biological organism—which, presumably means a human body with a functional circulatory system.

A more important problem is an iteration of the inference-defect. Just because someone has made a distinction between two things, one of which is valuable, it doesn't follow that the other thing is not valuable. Why think that our circulatory system is not valuable? Why think that my brain is not valuable? Certainly they constitute my biological life and I happen to think they are quite valuable! They are valuable not merely as enabling me to have good experiences. A functional brain is valuable, period. Having a functional thoracic system is valuable, period. One might say that they are only valuable in so far as they allow me to have conscious experiences. But why insert the term 'only'? Artistic talent is good even if the person who has it never makes use of it, i.e., it doesn't ever "enable" the person to sculpt or paint. My capacity to choose is a good thing, even if I make bad choices. My intelligence is good even if I come to believe false things (Gomez-Lobo & Keown, 2015). These are all cases in which a capacity, power, or functioning is good quite simply. Proponents of euthanasia are correct that just because something T enables or brings about an intrinsically valuable state of affairs does not entail that T itself is intrinsically valuable. My point here is only to point out that it does *not* follow that T is *not* itself intrinsically valuable either. Inferring that my biological life is not valuable simply because (a) one distinguishes between it and a biographical life and (b) supposes that one's biographical life is valuable, is an inference defect.

Another problem with the distinction is that on a straightforward reading of its proponents, it entails a body-self dualism (Lee & George, 2007). If it is understood this way, it falls prey to the concerns raised by many who deny dualism about human beings (Olson, 1997). To take just one of those concerns, imagine a psychiatrist who treats a person for multiple personality disorder. On some understandings of body-self dualism, the psychiatrist committed murder (Kaczor & George, 2017, 76). But that is deeply counterintuitive. The distinction should be understood, rather, as a distinction that is picking out two different aspects or features of one and the same human being; one of which is valuable, and the other is, supposedly, not.

Understood as preserving the unity of the human being, the distinction tells us that what is valuable is consciousness, or the fulfillment of one's interests, and the avoidance of pain. As has been pointed out above, P 300 event-related potentials are the putative neural marker for consciousness (Rutiku et al., 2016), as well as local gamma band responses (Aru & Bachmann, 2015), or the presence of functional synapses (McMahan, 2007), and that c-fiber firings are a marker for pain. But there is nothing morally significant about P 300 event-related potentials, or synapses, and so on. There is nothing morally significant about the neurochemistry that happens in the wake of accomplishing something great. These neuro-physiological states are not valuable in themselves, they are valuable insofar as they bring about or enable valuable states or experiences of the person. So, the distinction between biological and biographical life is a distinction along the same continuum. Take the entire conscious life of an individual complete with her experiences and fulfillments and contrast them with their neurological/biological counterparts. The biological processes are not valuable in themselves, the states they give rise to, however, are. This, I believe, is the most plausible way to understand the distinction.

Even so, the concern about artificiality and an inference defect persists. Is it obviously true that P 300 potentials have no value, or that c-fibers firing have no disvalue? Isn't this much like saying that H_2O has no value but water does; or that oxygen has no value but respiration does? The distinction is without a morally significant difference. Of course, in saying that one's biological life is valuable, it does not necessarily follow that one should keep it alive as long as possible. There are other values that might be relevant like avoiding disproportionately burdensome life-sustaining treatment. Withdrawing such support does not involve an intention to kill me (Tollefsen, 2006).

There are, however, at least two additional problems[3] with this distinction that should enjoy wide acceptance. The first problem I take to be decisive: it rules out our having bodily rights. The claim that being a "mere" biological organism bears no moral status or has no important moral properties denies that we have bodily rights. If human organisms are bodies and organisms bear no moral status, viz., no moral importance, and moral rights protect things with value, it follows that our bodies cannot bear moral rights. Raping an unconscious patient would not be a violation of her bodily rights being that she is a "mere biological organism." Since I am interpreting the distinction as preserving the unity of the human person, such violations of one's bodily rights are violations of the person. Even on constitutional accounts of the human person, damage to one's body should still be understood as morally charged. The typical examples used to illustrate constitution are things like

---

[3]A third problem is canvassed in the next chapter according to which, there are other aspects or features of our worth that do ground a prohibition against killing us even if we are suffering and request to be killed.

bronze statues. Such statues can be melted down into bronze—the bronze persists but the statue is destroyed. If the two things have different persistent conditions, they must be different—even when the bronze constituted the statue. I won't air my reservations with constitutional accounts here, but even on such accounts, taking a sledgehammer to a Rodin *damages the statue*. Likewise, violations of one's body are violations of that person.

Second, understood as preserving the unity of the person, the distinction cannot be used to justify killing the person. There are several options to consider. One might understand the distinction being used to argue against an absolute prohibition on killing innocent persons along the following lines: One should not kill innocent persons because their biological life is intrinsically valuable. Their biological life has no intrinsic value. Therefore, one may kill innocent persons. Using the distinction in this way, however, is fallacious. We learned to translate 'because' clauses, if implication is meant, as the antecedent of a conditional. So, the first premise should read as,

- If a person's biological life is intrinsically valuable, then we should not kill that person.

But when one denies the antecedent, it does not follow that the consequent can be denied. If I am in Chicago, I am in Illinois. If I am not in Chicago, it does not follow that I am not in Illinois.

Another way to render Rachel's use of the distinction between biographical and biological life is as an independent argument. One's biographical life is the only thing that is valuable about a person. Therefore, if one's biographical life is worthless, then the person's life is worthless. Ending worthless lives, all things considered,[4] does not violate any plausible ethical norm. Therefore, it is permissible to end a person's life that is worthless. Euthanasia ends worthless lives. Therefore, euthanasia is permissible.

The de jure defect here is that the distinction between biographical life and biological life does not itself argue that one's biographical life is the *only* value. It may be that one is more valuable than the other, but not that the other has no value. It could be that the two are equally valuable. It could also be that the distinction is not exhaustive—as will be argued in the next chapter. Briefly, I have a body, and I have a biography. But I cannot be identical to what I have. Consequently, I might not be worthless even if my biography is. Again, preserving the unity of the human person, what is killed is still *me*, my organism is not something *else*. To the extent that I should not be intentionally killed, the distinction does not obviate the general prohibition against killing persons. Rachels and others think that the 'spirit' of such a prohibition is to protect biographical lives that are going well. But the prohibition against killing is typically understood as protecting a person against killing *whether*

---

[4] One might demur here and say that we need to consult family members who might not want the person to be killed. But at this point in the argument, directed as it is to an audience of Westerners, the moral weight of the family's concerns are not overriding of the person's own wishes.

or not her life is going well. The POW has a miserable life but retains a right not to be killed, even if he asks for it.

To be clear, the two problems represent two different de jure defects. The first problem is an instance of an justification defect according to which the distinction is subject to countervailing reasons—the distinction entails that we do not have bodily rights. The second problem is an instance of an inference defect according to which the distinction still does not give one reason to think that it is permissible to kill *me*. We are not given a reason for thinking that the prohibition against killing innocent persons *only* applies to protecting their biographical lives.

6.2.2 Can Consent Turn Murder Into Mercy?

At issue with the previous defect is that the distinction between biological and biographical life is insufficient to permit the killing of me. But the intuition remains that if the biographical life is worthless, are we not permitted to end it? The next defect explores more precisely the role that consent plays in the argument for (pk). This step is important in so far as suffering refers to experiences of the person that are bad, but persons are not *identical* to those experiences even if they *have* such experiences. One can easily imagine that if one's experiences were different, that person would still exist. We still need an argument for why it is permissible *to kill this person*, not just why it might be permissible to end the person's experience of bad things, i.e. by palliative sedation.

The argument that provides the needed vinculum is to focus on consent. One can see how consent turns rape into love making and stealing into borrowing (Hurd, 1996). Having bad experiences can turn killing into compassion if the person *requests* to be killed. Consent, one could argue, turns murder into mercy in the setting of suffering. If consent has such transformative effects on other actions (e.g., intercourse), it is plausible to suppose that it has such transformative effects in changing unjustified killings into justified killings.

There are several reasons for thinking, however, that consent does not exert its typical magical effects on the issue of euthanasia. It is plausible to suppose that making one's own decisions is valuable. When we think of why, however, we might refer to the many choices we make in our lives, our friends, career, what to eat, what to see, what hobbies to pursue and so on. Call these intra-life choices. Assuming these are valuable, it does not follow, however, that choices to end one's life are also valuable. Contra-life choices concern having oneself killed. If one already thinks that such choices are valuable, then one is assuming (pk). If one thinks that contra-life choices are valuable because intra-life choices are, then the inference suffers from a non-sequitur because these choices concern different spheres of objects—how to live for intra-life choices, and to have oneself killed for contra life ones. So, consent to being killed cannot change unjustified killings into justified killings because the evidence or reasons for valuing it is culled from the value of intra-life choices.

A second reason[5] is that it is not obvious that because someone consents to x, x is thereby permissible to perform. Sumner, to his credit, discerns that such an inference is oblique when he observes that "my right to life is a claim-right in which my claim against you (and everyone else) that you do not kill me is identical to a duty imposed on you (and everyone else) not to kill me (Sumner, 2022, 85–86). The claim right is against *others* but does not necessarily apply to oneself. Sumner's point is that it is not enough to argue that suicide is permissible, the proponent of (pk) must also say that it is permissible for someone else to kill me. So it may be the case that a patient can waive her own rights, but *the doctor* should still not kill. We need an argument for why it is permissible for one person to kill another person.

Sumner thinks that such an argument hinges on the claim that one's right to life is inalienable and he does not think that such a right is inalienable.

There are other options to consider according to which even if an agent consents to x—thereby alienating his right—another agent doing x might still be immoral. Consider Graham Greene's story *The Tenth Man*. The book focuses on a French lawyer named Chavel who is captured by the Nazi's and finds himself in a concentration camp. Adding to his aggravation of having been thrown into a concentration camp is that he's in a block with other Frenchmen who were in a state of life lower than his. Greene paints Chavel in somewhat stereotypical ways in which Chavel, a successful lawyer, is haughty, prideful, and casts a supercilious glance on those he considers under him. His cell block receives a decimation order in which three of the 30 prisoners are to be executed and they have to choose among themselves which three. They cast lots and Chavel is one of the three.

The other two who are chosen accept their lot, but Chavel panics and offers his wealth and property to someone else's family who will take his place. The other prisoners view such an act as cowardly and Chavel a poltroon miser. As it happens, someone accepts the conditions, a man named Janvier who is suffering from tuberculosis. Janvier's mother and sister would inherit Chavel's property and wealth if Janvier takes Chavel's place on the firing line. A contract is drawn up and is witnessed by the other prisoners.

Janvier consents to the exchange, and even defends his decision when the others criticize him. He says, "'Why isn't it fair to let me do what I want? You'd all be rich men if you could, but you haven't the spunk" (Greene, 1985). Greene reserves the remainder of the book to assess Chavel's moral character and the effects of his actions. Carl Elliott comments on this case that "even though the deal was freely made, we know that Chavel was wrong to make it" (Elliott, 1995, 92). He was wrong to take advantage of Janvier's poor health, vulnerability, and selflessness for

---

[5] In between the first and second problem might be the problem that our choices are permissible or impermissible depending on their objects. Acts are moral not because they are chosen; rather they should be chosen (or not chosen) depending on whether they are moral (or immoral). Making good on this problem, however, assumes that death is not a benefit, or that the object chosen in euthanasia (death) is not itself a bad one. So, the moral dependency of our choices on the appropriateness of their objects is not itself a de jure defect of an argument for (pk), though it might be a de facto one. Below I explore a potential de jure defect in the notion of death counting as a benefit.

his mother and sister. The story brings into relief the distinction between *suffering* or undergoing harm on the one hand, and *inflicting* harm on the other. Though Janvier consented to x, it was wrong for Chavel to inflict x.

Our moral intuitions might be occluded or distracted by the context in which the story takes place, viz., a Nazi concentration camp. But once we see the pattern, other examples emerge that are much more closely aligned with the features recapitulated in euthanasia. Consider the Meiwes case. Alexander Meiwes posted an ad on an internet site devoted to cannibalism, saying "seeking well-built man, 18–30 years old, for slaughter" (Finn, 2003). Bernd Juergen Brandes accepted the invitation and after several email exchanges, Brandes visited Meiwes's home where the consensual killing took place. Initially Meiwes was convicted of manslaughter and not murder because, his defense argued, the victim consented. Germany's highest court eventually charged him with murder and sentenced Meiwes to life. Even if Brandes can alienate his right to life, it was still wrong *of Meiwes* to kill him.[6]

So, we have some cases in which consent exerts moral magic converting stealing into borrowing, and some cases in which consent doesn't change the nature of the moral action performed. Here is how to understand the differences. For each dyad, (e.g., theft vs. borrowing) consent is part of the definition of the action dyad. Theft is taking another's property *without the owner's permission*; borrowing is taking another's property *with the owner's permission*. Love-making is sexual intercourse *with the other's consent*; rape is sexual intercourse *without the other's consent*. Consent functions as an essential feature that identifies the action-types in question.

There are other action-types that can be specified without reference to consent. The Meiwes case suggests that murder is one of them; maiming, torture, enslavement, and possibly bullying are other examples. One may consent to being bullied but the act of bullying remains wrong and even when the consent is altruistic, for example, the inflicted is distracting the bully away from other people for whom the inflicted cares, e.g., a younger sister or brother. These reflections suggest that there are two classes of action in relation to consent. There are actions for which consent changes the action-type, as with theft and borrowing. And there are actions for which consent does not specify or define the action-type in question. Under this second category, there is a further subset for which consent does not *justify* another person doing the putative harmful action. (There is a distinction between defining an action and justifying it.) Consent does not change a morally unjustified action into a justified one, and the Meiwes case suggests that killing is just such an action that is recalcitrant to such a change.

To see this more clearly, consider the relations here stripped of any description of the act in question. Consider an agent (A) and a subject of A's action (S). S's consent does not entail that A must thereby respect S's end-setting capacities. One can easily imagine that S's consent does not entail that A no longer views S as an object. S's consent has no apparent causal link with how A *sees* S. Not knowing anything about

---

[6] Of course, the other conclusion to draw is that Brandes *cannot* alienate his right to life. But I am granting for the moment, Sumner's assumption that one can do so. The argument in the text is that it still does not follow that another agent is exculpated if he were to kill Brandes.

what kind of action is performed, we can see how S's consent does not tell us whether that consent changes the action-type of A's action. That act could still be an action of the type *using S as a means*, or as *harming S*; it is certainly logically possible that S consents to being used or to being harmed by another. So, S's consent alone does not morally justify A's action.

Applied to euthanasia, proponents of (pk) cannot use consent as a justifier for (pk). Killing is the kind of action that is recalcitrant to consent's moral magic. Of course, none of the comments in this section entail that (pk) is not justified. The only point here is that *consent* does not provide that justification.

What does the justifying work for (pk) is suffering and the related ideas of death being a benefit or life not being worth living. These more central issues suffer important de jure defects.

6.2.3 Death Is Not a Benefit

There are two problems under this heading. The first problem pertains to the argument that death can be a benefit. And the second is the problem of inferring from death being a benefit for S to it being permissible to intentionally kill S. I consider these problems in turn.

The Claim Itself

Sumner uses the cases of Anita and Bill to motivate the judgment that death would benefit them. But this claim conflates several things that might be beneficial for them. It is beneficial for Anita that her suffering ends. All parties to the debate agree with that statement. It does not follow from this claim that it is beneficial for Anita to be *killed*. Inferring from the claim that they would benefit by not suffering to the claim that euthanasia is permissible requires an intermediate step. At issue in the euthanasia debate is whether the action of intending to kill an innocent person in certain clinical circumstances is a justified exception to the general prohibition against killing innocent persons. Holding that there are no exceptions to this prohibition is consistent with the claim that suffering is bad and may permissibly be eliminated even if by, for example, terminal sedation.[7]

The meretricious nature of the inference from ending (intractable) suffering is good to death is a benefit, is revealed by understanding how states of affairs are comparatively evaluated. Suppose a state of affair, SOA_i is evil or bad. It is sensible to infer that if SOA_i is evil, then $\sim(SOA_i)$ is good. But in making this inference, *we have to keep the states of affairs constant*. If we do so, it follows only that not

---

[7] Sumner grants that terminal sedation is a palliative option for Anita and Bill, but does not appear to recognize the moral chasm this introduces as outlined in the text. He states.

suffering is good. *Not existing* is not the evil. It is existing *while experiencing suffering* that is evil. The phrase "ending" suffering, however, goes beyond a comparison between the value that states of affairs may have, and describes an action; a specific action that eliminates suffering by eliminating the sufferer. What is important to see here is that the sufferer need not have intrinsic worth as that is understood by proponents of ~(pk). It is enough to recognize that the inference from ~(SOA$_i$) is good, to killing is permissible commits a categorical error. One is acting to eliminate the wrong thing or at least a thing that is categorically different than what is agreed by all to be bad, namely suffering.

To see how disparate suffering and existing are consider what I refer to as the continuum argument from Kaczor and George (2017). Kaczor and George consider the claim that death can be a benefit. Imagine a sliding scale of functionality ranging from an Olympic athlete at the top end, with death at the bottom terminus. As one descends the scale of functionality, one loses more and more functionality, such as strength, endurance, immune response, and so on. Now imagine that an agent intentionally brings about in another agent a drop or descent in that person's functionality. Typically, such actions would be judged wrong (unless it is an amputation, for example, to preserve the patient's functionality and prevent death). They state,

> It is difficult to understand why somewhere along the scale of physical well-being intentionally killing human beings becomes not a harm inflicted upon them but, rather, a benefit to them…The further down the scale the individual goes, the worse off the individual is in terms of physical well-being. To kill an individual is to completely destroy the physical well-being of the individual (Kaczor & George, 2017, 72).

Here is a thought experiment where we are asked to envision a sliding scale of functionality, and that the terminus of one's functioning is death. The intuition is that descending further down the scale of functionality is clearly not a benefit. And, there is no reason to think that arriving at the scale's terminus is magically transformed into a benefit. A second intuition is that an agent who intentionally brings about in another agent a descent in that person's functionality would be doing something wrong. Again, this wrong doesn't magically change when one brings about the terminus of one's functionality, namely death. Call this the continuum argument.

Consider the continuum argument in reverse according to which we begin by considering what we mean by suffering. The reason why we say that someone is *suffering* is because they have *lost* health and functionality. They suffer because they are deprived of something, for example, immune function, renal function, etc. But one cannot infer from 'it is bad that they have lost such health and functionality,' to 'it is good that they lose all functionality.' Suppose instead that by suffering what we mean is that the agent is experiencing a frustration of his or her interests. Again, we cannot infer from 'it is bad that the agent's interests are being frustrated,' to, 'it is good to eliminate the possibility of having interests at all.' These inferences are clearly fallacious. So, as we descend the scale of functionality, arriving at death suggests that death cannot be a benefit.

Of course, the thought experiment requires understanding death along the same continuum as loss of one's functionality. Losing one's functionality along the scale is, we may suppose, contrary to the time-relative interests of the individual who is suffering such losses. With death, the reply to Kaczor and George might be, there are no interests that can be further frustrated. Death is the terminus of the scale in terms of functionality, but not the terminus in terms of one's interests being violated—because there are none. In a way, this is the ghost of the response to the impairment arguments encountered in Chap. 3. Death cannot be included on the continuum (understood morally) because with death there is no possibility for violating one's interests. Just further up the functionality scale, however, involves considerable violations of one's time-relative interests.

This response suffers the same fate as it suffered in Chap. 3, namely, it is viciously circular. It assumes that death itself is not a harm to the individual. One might think that such circularity is benign as there are good considerations in favor of thinking that death cannot harm the individual who dies in certain circumstances. To think that death itself is a harm, one might argue, must presuppose something like the view that Blanshard labels as Thomist. And the argument against it follows Blanshard's riposte: Consider a life with consciousness away, and all value vanishes with it. Here is Jonathan Glover commenting on the view that life (biological life) itself has intrinsic worth. "I have no way of refuting someone who holds that being alive, even though unconscious, is intrinsically valuable. But it is a view that will seem unattractive to those of us who, in our own case, see a life of permanent coma as in no way preferable to death" (Glover, 1977, 45). And Rachels, quoted above, offers a similar thought saying, "in the absence of a conscious life, it is of no consequence to the subject himself whether he lives or dies" (Rachels, 1986, 26). To illustrate the ubiquity of this sentiment, here is Alan Goldman: "It is plausible to maintain that life itself is not of intrinsic value, since surviving in an irreversible coma seems no better than death" (Goldman, 2010, 77). The point in these quotations is that at some point in the loss of one's functionality, death doesn't matter to the individual who loses such functionality. If it does not matter to the individual, death can't be a harm. From the perspective of the subject herself who envisions being permanently unconscious, life will not matter. Does this view argue that death can be a benefit?

The first observation to make is that even if correct, these comments only argue that death is a benefit for those in a permanently unconscious state. We might revise the comparisons and consider two states of affairs, one in which one is suffering intractably, and the other in which one does not exist. If there is no preference against being dead, then death is a benefit. But even this revision does not justify thinking that death is a benefit in such scenarios. One reason is that what is clearly preferred is that one *not suffer*, not that one not exist.

A second response has to do with the procedure used to justify that death can be a benefit. The procedure is to pick out a scenario in which one suffers significant harm (i.e., permanent unconsciousness), and then we add the *putative* harm of death. If death does not add any additional harm, as evidenced by preferring it to continued existence, one concludes that death is not a harm. From the supposition that one would not prefer continued existence, however, it does not follow that death is not a harm. It may be that both states of affairs are viewed as bad, but one is worse than the other. Imagine being in one of the World Trade Center tower's on 9/11 facing either being burned to death or dying quickly by hitting the ground. In preferring the latter, one is not preferring *being dead*, or *hitting the ground*. Furthermore, it doesn't follow that being dead or hitting the ground is not a harm simply because one prefers a means of dying that is quick and relatively painless. Preference is not a reliable indicator of what is valuable, certainly not in forced-choice situations involving a lesser of two evils.

A point that continues to reemerge is that what is preferred in such scenarios is that one not suffer, not that one not exist. To turn Blanshard's riposte around, one can imagine suffering away, and disvalue vanishes with it.

To close out this branch of the dialectic, the continuum argument has us understand suffering as involving the absence of functionality. Therefore, it doesn't follow that eliminating *more* functions would be a benefit. And since death is the terminus of one's functionality, death cannot be a benefit. If we understand suffering as a frustration of one's interests, it still doesn't follow that eliminating the possibility of taking an interest at all is a benefit. The standard arguments used to justify the claim that death is a benefit, such as preference type arguments, fail to justify such a claim.

The Inference

Now suppose that death is a benefit, that is, not only is not suffering a benefit but not even existing is also a benefit. It still does not follow that killing the person is permissible. Consider cases in which a person can benefit from another doing action A, but A still should not be done. A person might benefit (in some sense of benefit) from undergoing so-called Asian eye surgery, but it still should not be done (Little, 1998). Such an action accommodates or recapitulates Caucasian conceptions of aesthetic appearance and are arguably racist. In fact, many forms of cosmetic surgery fall into this category where the person might benefit from the surgery—in some way—but the surgeon should not do it.

Other examples include a black person taking pigment conditioners to lighten one's skin, or dyeing one's hair blond, and so on. Such actions evince racist norms and should not be done even if the person thinks it will benefit him or her. More generally, consider cases in which an injustice occurs but it is not understood as an injustice because of—if true—deeply ingrained prejudices secondary to capitalistic or neo-liberal society, e.g., instances of *systemic* racism. Other examples include conversion therapy for homosexuals in cases where the homosexual person requests it. Assume that such therapy would be 'effective' in some sense according to which the therapy meets *the agent's* own personal and therapeutic goals. Considered as such, it would benefit the person—in some sense—but many may think that the therapy should still not be done.

These examples might be cases in which the agent is confused about what would truly benefit him or her. This reply, however, is not available to any defender of euthanasia who puts a premium on the value of self-determination. These are cases in which *the subject* views death as a benefit, but the action that putatively benefits the subject should still not be done.

But what follows if we understand death as a benefit objectively? Suppose that the person is *correct* in thinking that death is a benefit for herself. Does that fact justify another person killing her? That argument might look something like the following:

  (i)   Death is a benefit for S in some circumstance C (where C = some state of intractable suffering).
 (ii)   Death is a benefit for S in C only if death is a means to end intractable suffering.
(iii)   If death is a means to end intractable suffering, then (pk).
 (iv)   Therefore, if death is a benefit for S in C, then (pk). (From ii & iii).
  (v)   Therefore, (pk). (From i & iv).

Premise (ii) is faithful with the traditional justifications for (pk). Authors typically envision cases of intense or intractable suffering in hopes that readers will resonate with the moral intuition that such suffering must be stopped somehow, even if by death. Proponents of (pk) may quibble that being a means to end intractable suffering is merely a necessary condition for making death a benefit. They may think that it is also a sufficient condition as well. I agree that it would count as a sufficient condition if the premise claimed that the *only* means to end intractable suffering is by being dead. If death is the only means, then it might count as a benefit objectively understood. If, as is actually the case, terminal sedation is a means to end the awareness or consciousness of suffering, then death is not the only means. Hence, (ii) is stated as a necessary condition.

In response, consider two different means to a good end—e.g., the good end being knowledge of vital importance on the toxicity of a new gene therapy drug. As a researcher I could either, administer the infusion at the same time across all subjects, or I could administer the infusion sequentially, thereby waiting for any immediate toxicities to manifest themselves before giving the infusion to the next subject. Clearly the latter means minimizes the risks to the subjects as a group and should be selected. The former scheme is still a means to obtain knowledge and should be considered as a necessary condition for counting as a benefit. But it is clearly not sufficient.

Likewise, for death to count as a benefit it is a necessary condition that it function as a means to end something that is unqualifiedly bad. Whether ending something that counts as unqualifiedly bad is a sufficient condition requires judging the different means in relation to one another. It might be that other means are better when measured by other relevant values at stake. More concretely stated, just as a sequential infusion minimizes risks but reaches the same beneficial end, so too suppressing one's consciousness of suffering reaches the end of ending suffering, but without the moral risk of an intentional killing.

One might reply and say that this objection to (ii) abandons the assumption that death *is* a benefit. The idea with this assumption is that if death is a benefit, it matters not how it relates to other means. The research example above imports the idea that one of the means invites greater probability for harm. Exported to defend (ii) however, this analogous feature must view death with moral suspicion.

In response, what needs defending is that (i) gives us a reason to accept (pk). It might be the case that being vaccinated is an objective benefit to someone, but it still may not be permissible to force vaccination upon someone who refuses (assume that others are not harmed by the refusal). The reason is that there are *other* values at stake that might be impugned by an action that brings about the benefit in question. That is why there needs to be a sub-argument for thinking that (i) gives us a reason for (pk) so that other values are not inadvertently impugned by an action that brings about death.

What other values might be at stake? One value might be parsimony in choosing the most efficient means to the good end. The value of parsimony might tells us that terminal sedation should be privileged because it ends the consciousness of suffering without also ending the very existence of the sufferer. The argument for this is more ecumenical than proponents of (pk) may think.

Here's how to understand the point so far: both the sequential and the non-sequential infusion schemes give you the beneficial end of knowing what the drug does to the person. Likewise, both terminal sedation and euthanasia give you the beneficial end of ending suffering. But only the former avoids the moral risk of killing. Death is a benefit only in cases in which it ends suffering. But because suffering can be ended without killing, it does not follow that euthanasia is permissible.

If death is a benefit according to (i), it is so *only* in very localized circumstances—recall Lach's commentary on the terminally ill patient versus the love lost teenager. The claim that killing is permissible because death is a benefit bears a burden of proof, the presumption is that killing an innocent is not permissible. We need a reason *for* thinking that death is a benefit and that it justifies killing. One reason is that it ends an SOA of disvalue. But other means do this too, as mentioned, terminal sedation. In fact, *there is no corresponding presumption against ending the consciousness of suffering as terminal sedation does*. So, the value of parsimony is a value that gives one reason for privileging a means to end suffering that does not entail (pk). And this is for reasons with which proponents should hold sympathy.

These comments merely set up the real problem with the argument, which concerns premise (iii). Why think that death being a means to end suffering justifies killing the sufferer? The problem here is not moral, it is conceptual. What is good is that suffering is ended. Suffering being bad or evil, however, does not entail that being dead is good. And just because death ends something bad, it does not follow that the *means* of killing the sufferer is good—consider the sufferer's arch enemy satisfying his request for euthanasia. From the supposition that ending the consciousness of suffering is good, it does not follow that actions that bring about death are also good.

Of course, the doctor performing euthanasia does not harbor malicious motives as with the arch enemy. But other admittedly more subtle factors can pollute an action that otherwise brings about a benefit. Research on prisoners, high risk research on the terminally ill, and phase I toxicity trials on the poor, the homeless, or racial minorities (Elliott, 2010) are all morally problematic in ways analogous to the problems with (iii). The moral problem is not that the subjects cannot give informed consent. The moral problem is not that such research can bring about benefits. The moral problem is that such research may *take advantage* of the subjects' vulnerability according to which the subjects would not enter such research had it not been for their dire clinical or social economic status. Likewise, patients who request euthanasia would not otherwise request it were it not for their dire clinical circumstances according to which they are suffering. The doctor would be following through on choices, even informed choices; but these are choices that they would otherwise not make had it not been for their suffering and vulnerability. Intractable suffering places duress on the sufferer and it exerts morally analogous concerns as are raised in the context of research on prisoners, high risk research on those who are clinically at a dead-end and so on. So, even if death may benefit S in circumstance C, it does not necessarily follow that killing S is permissible since such an action might take advantage of or exploit S's vulnerability. Since there is no presumption against using terminal sedation in these circumstances but there is for intentional killing, the argument for (pk) is unsupported.

6.2.4 Suffering Is Not a Feature of Persons That Justifies Killing Them

The epistemic bar the argument for (pk) must surmount is that there is a justified exception to the moral prohibition against killing innocent persons. What justifies an exception to this prohibition if anything? We have discovered so far in the review of previous de jure defects that the value of autonomy or self-determination is not sufficient (Conly, 2013). Is it jointly sufficient with the added feature of suffering?

Suffering is not the kind of thing, feature, or SOA that justifies killing the sufferer. Consider other cases of suffering according to which one is not justified in killing the sufferer, the POW, the love lost teenager, the grief stricken widow, and so on. Lachs seems to think that persons with average moral intelligence will just see the difference between the stage 4 cancer patient, for example, and the love lost teenager. There certainly is a difference, but not one that justifies killing one but not the other. One could just as easily say that neither should be killed. Lachs is assuming that it is permissible to kill *in just* the circumstances in which proponents of (pk) want it permitted. But there is no reason offered for this other than 'look and see.'

6.2.5   *The Groundless Objection*

This leads to the final de jure defect. The key premise in the argument for euthanasia is premise one which reads,

(1) It is permissible to kill an innocent person if and only if (i) the person requests to be killed and (ii) is suffering harm to the degree and kind that the person judges unbearable.

This section argues that the last de jure defect is that the argument for (pk) must be viciously circular.

Consider other cases of suffering to see if suffering in those contexts provide intuitive support for (1) such as killing the POW, the concentration camp victim, the severely depressed patient, or the lovelorn teenager, all of whom might be suffering and they might want to die (call these "the alternatives").

Either killing the alternatives is permissible or it is not. If not, a proponent of euthanasia has an adjustment decision to make. She may reject the permissibility of euthanasia or not. If she continues to accept (pk), she must find a feature F that is present in the alternatives but is not present for those she thinks are in euthanasia-permitting circumstances (or vice versa). Justifying this distinction is to ground a reason for dividing the class of innocents into those for whom it is permissible to kill, and those for whom it is not. How is one to argue that (i) there is a feature F, and (ii) it is a morally relevant one?

Before answering this question, it is necessary to understand an elementary epistemological distinction introduced in Chap. 1. Beliefs are divided broadly into basic beliefs and non-basic beliefs.[8] Non-basic beliefs are justified on the basis of other beliefs. Basic beliefs can either be justified or not, but not based on other beliefs. If they are justified, then they are justified based on non-doxastic states, such as an experience or intuition. Whether the experience or intuition is veridical, e.g., produced by a reliable faculty, will dictate whether the non-doxastic state in question can justify the basic belief. Though the following examples are controversial, they illustrate the point behind this technical language. Perceptual beliefs based on visual sensing at medium distances in good lighting are examples of justified basic beliefs since they are justified by perceptual experiences which are not other beliefs. My belief that the piano I see has less than 88 keys is justified, however, on another belief, namely, that I am at a concert purporting to use period instruments.

We can now answer what justifies the distinction between killing a terminally ill patient and killing one of the alternatives. Either the distinction is a basic belief or it is inferred from another belief. If the distinction is basic, this would amount to saying that it is permissible to kill the terminally ill patient, but not permissible to kill the concentration camp victim or the POW, full stop, no reason. If the distinction is a basic belief, (1) *just is* true with an additional conjunct that killing the alternatives is impermissible. This strategy is clearly unsatisfactory.

---

[8] See Audi (2001, 32 ff.).

If it is a non-basic belief, it must be justified on the basis of other beliefs. What other beliefs? One feature that is different between the two classes of innocents might be that the cause of suffering for the terminally ill patient is internal (i.e., disease), and for the POW it is external (i.e., captors). Why think that the internal/external distinction provides a sufficient reason for distinguishing between permissible and impermissible killing? Again, this is either a basic belief or a non-basic one. If it is basic, it is wholly unconvincing. There is nothing apparent about causes of suffering that are internal that would justify killing me; likewise, causes of suffering that are external is hardly *the* reason for not killing me. If it is non-basic, it must be justified by another belief. Again, we can ask what other beliefs? If that other belief just is the conjunction of (1) with the belief that it is impermissible to kill the alternatives, this would be circular reasoning since what needs defending is premise (1).

Must the defense of (1) be justified by circular reasoning? Return for a moment to the general prohibition on killing, namely, that it is impermissible to intentionally kill an innocent person. That S is innocent is either a sufficient reason for making it impermissible to kill S or it is not. If it is a sufficient reason, then ~(pk) follows. If it is not a sufficient reason, defending *that* claim must think that it is permissible to kill *some* innocent persons. And that claim is entailed by (pk). The conclusion, (pk), would have to be used or assumed in defense of (1).

Here's another way to approach the circularity argument. The two conditions which render euthanizing S permissible are that (i) S has requested it, and (ii) S's suffering is intractable. If both conditions are met, then (pk) follows. But what if S being innocent is a sufficient condition for making it wrong to intentionally kill S? Or, what if the value of parsimony is respected only by utilizing terminal sedation? If either option were true, it would block the inference to (pk).

The argument that the innocence condition, for instance, is not sufficient would have to be argued for by taking a stand on cases in which one thinks that killing the innocent is permissible. And to do that, one has to assume (pk). There might be cases such as the spelunker case in which it is permissible to kill the corpulent man who is blocking the only route of escape for the trapped miners. Of course, it wouldn't follow from 'killing the corpulent man is permissible' to 'euthanasia is permissible'. It matters what feature(s) is making it permissible; and there is no reason in the spelunker case to say that such features are duplicated in the euthanasia case—the most obvious of which is that the corpulent man is *not* suffering.

One can see a pattern developing here and prefigured in Chap. 3. Counterexamples are presented according to which a person is suffering, and requests to be killed, and yet it is still wrong, presumably, to kill the person (e.gs., POWs, the depressed lovelorn teenage, etc.). Deflecting the epistemic pressure of the counterexamples must appeal to (pk) itself.

Likewise for the argument from innocence. What is the reason for *precluding* innocence as a sufficient condition for making it wrong to kill S? Such an argument has to presuppose that there are some innocents for whom it is permissible to kill.

Which ones? Whereas it might be the case that some circumstances are such that it is permissible to intentionally kill an innocent[9] (I do not think that there are any, and the spelunker case is not a case analogous at all with euthanasia nor is it a case of intentional killing), it would have to be argued that there are such cases. Citing the euthanasia case as an instance is obviously circular.

Engaging in reflective equilibrium across different cases is not promising either. Judicial murder cases used to motivate utilitarianism are cases in which the murdered person is not suffering. The POW case is one in which the prisoner is suffering and might even request to be killed. On this point one has an adjustment decision to make. Is it permissible to kill the prisoner in such a circumstance? If the intuition is no, then the POW case cannot be used to motivate permissible killing in the context of euthanasia. If the intuition is yes, one already thinks that it is permissible to kill innocent persons in certain circumstances. Only those who already have such an intuition would think that such a procedure justifies (pk).

Consider the depressed teenager who is unresponsive to therapy and continues to request euthanasia. Some think his request may be respected. We want to know why. The point I want to make is that the only explanation is that one believes that it is permissible to kill an innocent person if the person's suffering is intractable. That is, one assumes (1) when what we wanted was a defense of (1). This way of proceeding can be iterated across all cases that are viewed as relevant. Either the case invites the intuition that it is permissible to kill the innocent person or it does not. If it does not, it can't be used to motivate any intuitions regarding euthanasia. If it does, we can ask why the person has such an intuition. The only answer that can be given is that the agent thinks it is permissible to kill innocent persons who are suffering intractably and request to be killed. (Pk) is assumed.

6.3 Conclusion

In summary, one can see a few principal themes emerging in each of these defects. One core theme is the notion of a categorical error one iteration of which is that the argument for (pk) moves from suffering being bad to the sufferer being worthless; or the move from the sufferer being worthless to a permission to kill the sufferer. I do not discern a reason to think that suffering is a reason to kill someone. Another theme recapitulates points introduced by Longino (1990) in Chap. 2, according to which (pk) itself functions as a background assumption.

There are some promising and enticing thoughts in the argument for (pk). But when we interrogate them we find that they do not even leave the driveway. The distinction between biological and biographical life looks so promising, but there is nothing in the distinction itself that suggests that one's biological life is *not*

---

[9] I do not think that there are any such cases. The spelunker case is not a case analogous at all with euthanasia nor is it a case of intentional killing, though I recognize that the latter claim is more controversial.

valuable. The claim that death is a benefit or that life is not worth living look enticing, but they too are meretricious due to the continuity argument. Furthermore, making good on these phrases requires comparing two SOAs one of which lacks the item being compared, i.e., the subject. In each case, we have a non-sequitur, we lack a reason to think that there is an exception to the prohibition on intentionally killing innocents.

Chapter 7
Euthanasia, Suffering, and Care

7.1 Introduction

The previous chapter addressed the principal argument for the permissibility of euthanasia (pk) and proceeded to canvas five de jure defects of the principal argument. The guiding ideas governing those defects are that suffering and respect for autonomy do not suffice to suspend or exempt the prohibition on intentionally killing human persons. Features of a person's experience (i.e., suffering) and state of will do not justify an exception to the rule against intentionally killing that person. The principal reason why is because though suffering is bad, the person who is suffering is never bad, or worthless, or entirely lacks value. The suffering and the sufferer are different and either can possess incommensurable value (or disvalue). In the setting of a presumption against intentional killing, which creates a corresponding burden of proof,[1] the argument for (pk) is unsupported or under motivated.

This chapter explores the merits of an argument for not-(pk), hereafter ~(pk). There are typically two ways of arguing for ~(pk). The first way is more abstract in that it focuses on the *type* of action that euthanasia is namely, the intentional killing of an innocent patient. And it argues that all such actions are contrary to justice and therefore impermissible.

The basic structure of those arguments is to argue first that you and I have *intrinsic* dignity which means one of the following ideas: it means that we have a worth or value that does not go away simply because we have a disease or are suffering. Our worth is not exclusively a function of how good our experiences are or how functional we happen to be, although good experiences and being functional are valuable. There is an additional value to who we are as human persons, and that

---

[1] Presumptions and the burden of proof are corollaries of one another. A person who has been missing for 7 years or more is presumed dead. Those who wish to argue that the person is alive bear a burden of proof. A person standing trial is presumed innocent, those arguing for his guilt bear the burden of proof. For whatever is presumed, it's negation bears the burden of proof (Freeman, 2005).

S. Napier, *Justified Killing*, https://doi.org/10.1007/978-3-032-14946-6_7

value does not vary or waiver when we get sick. Gómez-Lobo and Keown (2015) observe that just as eyesight is good even if we come to see bad things, so our life is good even if we come to experience bad things. Just as intelligence is good even if we come to believe false things, so too our life is good even if one is disabled or ill. Some authors use the language of rights and argue that our right not to be intentionally killed is inalienable. The person herself cannot alienate or waive her right not to be intentionally killed. I would argue that the focus on rights is parasitic on the former approach that focuses on the intrinsic dignity of the person. The reason why we cannot alienate or waive our right to life is because it is a right the kind of which is not up to us to discharge. We have a worth that is independent of our will or our own assessment of our worth. Again, Keown observes (2018) that we have inherent worth even if we happen to lose sight of that worth.

The second step for such arguments is that it is a violation of justice to destroy something that has intrinsic dignity, e.g., human persons. This premise is more ecumenical than may initially appear. Even proponents of (pk) assume something like it in so far as they think that it is wrong to frustrate or violate someone's time relative interests, or to end a life that has a good quality to it. Frustrating, destroying, or violating something that has value, whatever that something is, is immoral. The difference between the justifications for (pk) and ~(pk) come to whether there is something valuable about the person *in addition to* the person's interests or quality of life.[2]

A second way of arguing for ~(pk) is to explore what the virtues of love or caring look like in clinical settings in which the patient is requesting to be killed. How should one care for such patients? What does loving such a patient look like? Proponents of (pk) exploit a similar strategy vis-à-vis the notion of compassion, but rarely is it argued that intentionally killing an innocent patient is compassionate. Proponents seem content noting that euthanizing such patients respects their autonomy and is therefore compassionate. The previous chapter addressed the de jure defects of focusing on the value of autonomy or self-determination to justify killing patients. Others seem content noting that the compassionate response to suffering is to eliminate it even if that means eliminating the patient. Kohl seems to think that not doing so must think that suffering is a good thing. Such a claim must conflate the person with the person's experiences. (For what it is worth, I find that killing patients who are suffering is far from a compassionate response but rather abandons them. It tells the patient that all she or he is, is their suffering and that they are not

---

[2] The only complication here is Sumner's (2011, §4.1) claim that even if there is something *intrinsic* about our worth, it matters *how much* that worth is, and whether it always outweighs the disvalue of intractable suffering. This is a hard comment to reply to in so far as if the proponents of ~(pk) are correct, the value of intrinsic dignity would be a different kind of value and therefore, incommensurable with the value of bad experiences. One could not justify that it outweighs the disvalue of suffering; but from this inability to do so it does not follow that suffering justifies killing an innocent patient for the simple reason that there is an additional value that is impugned, destroyed, or violated.

persons with irreplaceable worth whose experiences may be excruciating or horrific.)

For those who approach the issue of euthanasia through the lens of what certain virtues look like in certain clinical encounters, they need to provide a broader account of what the virtue in question means, and how that virtue principally operates. One should argue for and not just assume that the virtue's characteristic operation would dictate behaving in a certain way.

The plausibility of a moral argument depends in large amount on the moral sensitivities of the argument's audience. Cases, thought experiments, and narratives can be used to highlight one's moral proclivities for the purposes of showing that a particular premise in a moral argument is more plausible than its denial.

With these observations in mind, there are different ways to access or approach the plausibility of the principal moral claims that argue for ~(pk). The presentation here follows the two approaches just canvassed. For the first approach, I argue that we can access or 'see' the inherent dignity of the person through, so to speak, the emotions of remorse, grief, and love (Sect. 7.2). For the second approach, I articulate reasons for why the virtues of loving and caring for patients who request to be killed *preclude* intentionally killing them (Sect. 7.3). One might think that the latter route is parasitic on the former insofar as loving something that is worthless is not love, but misguided affection. But the focus of the latter route is to highlight one's moral intuitions on what love looks like, intuitions that might supervene on but are not identical to beliefs about the inherent worth of the beloved. Famously, Hursthouse comments on the issue of abortion that, "in exercising a moral right I can do something cruel, or callous, or selfish..." (Hursthouse, 1991, 235). Her point is that even if abortion were a moral right, one might still act viciously in exercising that right. She explains further, "Love and friendship do not survive their parties' constantly insisting on their rights, nor do people live well when they think that getting what they have a right to is of preeminent importance..." (Hursthouse, 1991, 235). The starting point for defending the second approach asks what would a loving and caring clinician think, feel and do for a patient whose suffering prompts him or her to request to be killed? Our moral intuitions on what love and care look like are predominant. Understood as such, the second approach can be motivated quite independently of whether patients have intrinsic dignity as the first approach requires. Both approaches are canvassed in this chapter to capture a broader range of intuitions and sympathies.

7.2 Human Dignity and the Importance of Intentions

Under this heading, there are two central components to the argument for ~(pk). What is argued is that a particular kind of human action, one whose aim is to eliminate the existence of a person, is impermissible. Second, the value or good that is impugned or eliminated by such an action is the intrinsic worth of the person. Even if the person's experience of suffering is ended, (a good thing), or if the person requests to be killed, the intentional killing would still violate or eliminate the inherent worth of the person. In so far as the action aims to destroy that which has inherent worth, the action is wrong on that count alone. So, a brief discussion of human dignity, or the inherent worth of the person, and the moral importance of intentions, is necessary to understand this argument for ~(pk).

7.2.1 Human Dignity

The phrase human dignity in contemporary bioethics is fraught with competing conceptions, different understandings of its relative axiological weight, and different purposes to which the concept is used in ethical arguments. Rather than canvas each view and subject it to detailed scrutiny, I focus on my own view which I have argued elsewhere (Napier, 2020) is quite ecumenical.

Most commentators agree that there are two different aspects of a human person's worth: human persons are *equal* in worth and we have *intrinsic* worth. Regarding equality, we typically assume that human persons have equal worth that is not dependent upon our social role, economic status, race, gender, political identity, etc. What grounds or makes us equal would be something that we share equally. So, our equality might be a function of shared properties or features such as the fact that all human persons have human nature, or are rational substances, or whatever your preferred view of persons is. Conversely, intrinsic worth is the value we have qua individual. Zagzebski airs a common moral intuition to the effect that "things with dignity cannot be compared in value to anything else,…That means we can never make up for the loss of the thing with dignity by replacing it with another or even many others" (Zagzebski, 2001, 402). Whereas a person with human nature, a good humor, or adroit intelligence, etc. are features or properties that are interchangeable and replaceable, you and I are not. Persons have irreplaceable worth that must be grounded in or a function of properties are features, if I can use such terms, that are incommunicable. Again, Zagzebski enlightens saying, ""incommunicability" is the name of a way of being that is unique to a particular individual. It is not shareable even in principle, and since qualities are shareable, it is non-qualitative" (Zagzebski, 2001, 415). This is not to say simply that being the only thing in the universe makes something valuable. A fingerprint might be unique, but a fingerprint as such is not valuable. The claim is, rather, that persons have irreplaceable worth not in virtue of the fact of being unique, but in virtue of the fact of being a person.

I assume that persons have dignity in the sense of being equal in worth. Less popular, but still, if I may, equally plausible is the claim that our dignity includes having irreplaceable worth. The best way to motivate this view is to see it through various emotive states since emotions are perceptions of value (Zagzebski, 2017), in particular, remorse, grief, and love.

An agent feels remorse when the agent believes that she has wronged another. Remorse is related to grief in that grief is the feeling of having lost someone or something of importance. Remorse is specific to the 'loss' of one's moral identity.[3] When an agent feels remorse, Raimond Gaita (2004) argues, the emotion shines a light onto the irreplaceable worth of the one who has been wronged. We can see this in a backhanded way by first considering whether the typical explanations for wrongdoing in modern moral philosophy capture precisely what the remorseful person responds to or sees. We cannot imagine the remorseful person responding, without parody,[4] as "What have I done? I have violated the social compact, agreed behind a veil of ignorance." "What have I done? I ruined my best chances of flourishing." "What have I done? I have violated the rational nature in another." "What have I done? I have violated the ideal interests of an autonomous agent." Such responses border on parody if they are meant to account for the content of a person's remorse. The remorse or sadness over wrongdoing is not explained by any of the typical explanations for wrongdoing. Rather, one is sad having offended, disrespected, or annihilated this person. The remorseful person comes to have a lucid grasp of having wronged *a person*. And what a remorseful person understands as wrong need not be reducible to harm, or a violation of one's considered interests, but could involve, as with a repentant arrogant person, a lucid realization that others are fundamentally equal to oneself.

Gaita's point in analyzing remorse is to bring into relief the fact that various accounts of wrongdoing on offer in contemporary moral theory border on parody. The principal reason why is because "the individual who has been wronged and who haunts the wrongdoer in his remorse has disappeared from sight" (2004, xxii). The preciousness of the individual is lit up by the remorse of having wronged him or her. The argument here is that remorse is a common emotion that is apposite in many cases. If the object of remorse turns out to be the preciousness of the person, remorse functions as a periscope by which one glimpses this aspect of human worth.[5] We might be submerged in a theory of rights, or in a time-relative interest account of harm, but remorse lights up the worth of this human being in ways untouched by these theories. A repentant murderer exclaiming "What have I done? I have killed something with psychological unity relations" hardly captures the content of her remorse. Consider also, "what have I done? I have killed a sentient being capable of forming and pursuing long-term projects and commitments" (Brewer, 2009, 172). These explanations border on parody.

---

[3] Ordinary usage of this term may tolerate a broader application, but it would not preclude the characterization given in the text.

[4] The following comments are adaptations or additions to Gaita (2004, p. xxi).

[5] In exploiting the metaphor of a periscope, I am suggesting two ways of considering the epistemic effects of remorse or other emotions. It might be that one is submerged, so to speak, in a theory and the remorse allows one to glimpse the moral reality which the theory had occluded. Or, if moral perception is theory laden we might say that the remorse introduced a new datum or intuition that does not sit well with one's current theories or beliefs (on analogy with anomalies as understood by Kuhn (1996, ch. 6)). Therefore, one must bring about a new and different "reflective equilibrium" that takes account of the new datum. Whatever we may assume to be the background epistemic position, remorse, grief and love have the same epistemic effect. Thanks to Ben Richards for pointing out these different interpretations.

A potential contender to Gaita's reflections might be interest accounts of harm. As Beauchamp and Childress (2001, 148) observe, "These arguments [for what counts as wrongdoing] suggest that causing a person's death is morally wrong, when it is wrong, because an unauthorized intervention thwarted or set back a person's interests…" At issue is not whether we feel remorse. The challenge is to argue that the remorse or regret is a function not of some nebulous preciousness of the person, but simply a function of having violated the person's interests.

In reply, the interest accounts of harm and wrongdoing fail to avoid parody or are simply counterintuitive. Consider friendship or parenting in which the growth of the friendship or parent-child relationship requires, at times, confronting and even thwarting the interests of the friend/child. When we do not try to dissuade a loved one from a wrongheaded decision and the friend/child is injured or significantly harmed from it, we feel apposite remorse precisely because we did acquiesce to their interests.[6] In such a case remorse cannot be understood as lighting up the value of someone's interests—quite the opposite. The only way to avoid such a result is to argue that the friend or parent in such cases ought not to feel remorse. That would be a counterintuitive position. Again, consider a penitent murderer who appreciates the wrongness of her actions. If the object of her remorse is expressible as having countervailed the interests of another, she has missed the moral reality of her action. Whereas, if she understands her actions as having killed *him*, "I have eliminated *his* existence" we should be content with her response as being accordant to the gravity of her offense. So, in many cases, remorse lights up the individual preciousness of the one wronged.

Consider next, grief. Brewer (2009, 174) observes that "mature grief at the death of a loved one involves an awareness,…that nothing could represent a compensation for what has been lost. Consolation might be possible, but compensation is not." We do not mourn the loss of a person's capacities, properties, or features. A person might be risible, gregarious, and affable, but we do not mourn the loss of risibility, gregariousness, and affability. Nor is it necessarily the case that we love the person because of those features. "But grief is not just a generic pro attitude towards an irretrievable entity with a certain set of *natural properties*; grief lights up its lost object as having had a very particular sort of value" (2009, 176, emphasis mine). Brewer's point is that the object of grief cannot be a property or feature of the person, but must be the person herself. Grief presupposes that something of value has been lost, not just the absence of good experiences. It follows that grief lights up the individual preciousness of the person.

Finally, consider love. When someone loves me, he does not love a feature or a property of me, but he loves *me*. Gaita recounts a story from Primo Levi who relates the story of his last days in Auschwitz when Russian artillery rounds could be heard in the distance. After years of suffering, liberation was a few weeks away. One of the

---

[6] Ideal interests do not help since the one who is wronged has then "disappeared from sight" since ideal interests might not be held by the person wrong, or by anyone for that matter, as they are not necessarily occurrent interests.

prisoners below his bunk was named Ladmaker. Ladmaker was a 17-year-old Dutch Jew who suffered from typhus and scarlet fever in succession, and a cardiac anomaly was developing. Additionally, because of his sickness and malnutrition he was bedbound and formed bedsores such that he could only lie on his stomach. He was always hungry, in spite of his fevers, and no one in his compound could understand Dutch which made caring for him difficult. One night, Ladmaker crawled out of his bed in an attempt to make it to the latrine. He was so weak, he fell to the ground sobbing in despair and pain. Levi then tells us how a bunkmate named Charles responded.

> Charles lit the lamp…and we were able to ascertain the gravity of the incident. The boy's bed and the floor were filthy. The smell in the small area was rapidly becoming unsupportable. We had but a minimum supply of water and neither blankets nor straw mattresses to spare. And the poor wretch, suffering from typhus, formed a terrible source of infection, while he certainly could not be left all night to groan and shiver in the cold in the middle of the filth.

> Charles climbed down from his bed and dressed in silence. While I held the lamp, he cut all the dirty patches from the straw mattress and the blankets with a knife. He lifted Ladmaker from the ground with the tenderness of a mother, cleaned him as best as possible with straw taken from the mattress and lifted him into the remade bed in the only position in which the unfortunate fellow could lie. He scraped the floor with a scrap of tin plate, diluted a little chloramine and finally spread disinfectant over everything, including himself (Gaita, 2004, xvi).

Gaita observes that Charles's actions could be classified as supererogatory, which they were. But the goodness of the action is not exhausted by the concept of duty, or supererogation. It is not exhausted either by the consequences of the action, namely, Ladmaker's comfort to the extent that he could possibly be comforted. An SS officer could have provided such comfort but still think all along that Ladmaker should die in the gas chambers. What Charles did was act *tenderly* towards another human being in the midst of extreme hardship and suffering. "Goodness, wonder, purity, love" (Gaita, 2004, xvii) are concepts that better capture the moral landscape than concepts that describe Charles's actions as beyond duty, or results in good outcomes, or satisfies Ladmaker's time-relative interests. "The wonder of what Charles did is that he responded fully to Ladmaker's degradation,…*while affirming Ladmaker's undiminished humanity*" (Gaita, 2004, xix, emphasis mine). Narratives like the one Levi hands down to us invite us to *see* through affliction and disability to the irreplaceable worth of the sufferer. What did Charles see? To what did he respond? A plausible explanation is that he saw clearly that the suffering of Ladmaker was bad; but with greater acuity he saw that Ladmaker retained inherent worth. Speaking specifically about affliction, Simone Weil makes a similar observation; seeing the other's value—to love the afflicted *without condescension*—is a "miracle greater than walking on water" (Weil, 1968, 172). Weil and Gaita are urging us to look more closely at the *irreplaceable* person as being animated with inherent value while at the same time immersed in affliction or disability.

What is it to condescend? Loving the afflicted without condescension involves apprehending the inherent worth of the person despite the brumous effects of the person's sufferings or "low quality of life" as we say. Conversely, to condescend involves the following.

> To look on a life as one in which it is unintelligible that there should be meaning is to see it as empty of what is distinctively human. It is worse than merely to see it as empty of goods and opportunities… If we find it unintelligible that anything could matter to someone living such a life, then we cannot think that any evil done to him, or by him, can go deeper with him (Gaita, 2004, 194).

This is not to say that condescension is not understandable. In response to patients experiencing terminal delirium, for example, it is tempting to see *nothing but* their affliction. However, like Ladmaker, and many others, there exists some*one* with undiminished humanity.

The reason this insight is a "miracle" for Weil is because it requires such perspicuous moral acuity. Loving the afflicted and disabled without condescension involves not having the thought that the afflicted would be better off dead or not having been born at all. Reflecting on Saint Theresa of Calcutta's work, Gaita notes that her compassion "expressed the denial that affliction could…make a person's life worthless" (Gaita, 2004, 202).

Such a love without condescension invokes wonder for Gaita and is miraculous for Weil not merely because the *actions* are meritorious, supererogatory, admirable, or that they issue from good motives or intentions. "The wonder which is in response to her [Saint Theresa] is not a wonder at her, but a wonder *that human life could be as her love revealed it to be*" (Gaita, 2004, 205). What Charles's actions and Saint Theresa's life advert us to is just how valuable those persons are about whom their loving activities were oriented. There is an adaequation between Saint Theresa's love for the afflicted and the beloved's preciousness. Her love reveals to us what might have been obfuscated otherwise, namely, the depth and comprehensive value 'still remaining,' if you will, in the afflicted and disabled. Saint Theresa's love reveal that her patients were "fully our equals" (Gaita, 2004, xiii). Saintly love performs a revelatory role. Certainly, saintly love provides fodder for philosophical reflection on moral heroism and supererogatory action. But it would be cutting the inquiry short if we stopped there. Saintly love is meant to sharpen our own moral acuity. "Sometimes we see that something is precious only in the light of someone's love for it" (Gaita, 2004, xxiv). Recounting stories of love is primarily meant to have us see more lucidly what the lover sees.

What is it, then, that has dignity? Quite simply, you and I do. Endorsing a similar view but defending it by other means, Patrick Lee and Robert George remark that, "all human beings have real dignity simply because they are persons" (2008, 411). So long as we exist, we are precious, irreplaceable, inherently valuable, and equally worthwhile. What then is the relationship between dignity and our existence? Following Gloria Zuniga, I hold that dignity and the person are a two item *Sachverhalte*. Zuniga, following Wittgenstein, "describes *Sachverhalte* as thinkable configurations of objects that stand in a determinate relation to each other…A speck must have some color, a tone some pitch, and an object of the sense of touch some hardness" (Zuniga, 2004, 122). Dignity is in the person as pitch is in a tone or hardness in an object of touch. Our inherent worth may be absent only if we do not exist. Though the object and its hardness are not identical, so too having worth is not identical to existing—the two are co-extensive but not co-intensive. One cannot be found without the other. Certain valuable functions, such as reasoning, and taking an interest in certain projects and plans, are valuable. And those who are able to engage in such functioning are certainly valuable. But the dignity of the person cannot be localized to actualizing a function or property. Dignity is, rather, a *dimension* of the individual person.

We should be careful in saying however that you and I have inherent dignity *because* we are persons. There are at least three ways to understand the 'because' in such a claim. First, 'because' can be understood as referencing a cause, as in "Joe hit a homerun because he hit the ball hard." 'Because' may also reference an explanation as in "Neo took the red pill because he wanted to get out of the Matrix." Finally, 'because' may delineate a more comprehensive understanding of a thing as when we say that this child will develop functional rational capacities because it is a human being; this piece of metal is gold because it has atomic number 79; and x boils at 212 degrees at sea level because it is H_2O. The relationship between, for example, having atomic number 79 and being gold is entirely unlike the relationship between hitting a ball and scoring a home run. One might say that discourse on moral status uses 'because' in the first or second senses as in "this person has moral status because she takes an interest in continued living." Having an interest causes or explains why she has moral status. On my view, dignity is something that we have because of what we are, not because of what we can do.

Arguing for this claim cannot be done by asking what is it that "makes" us have dignity since doing so already assumes that you and I could exist without it; such a question assumes a bifurcation between properties or functions that are valuable on the one hand and persons on the other. The argument for my claim proceeds by tuning our moral antenna to the moral frequencies, if you will, that dignity sends out. Such tuning requires recapitulating narratives and reflecting on certain emotive states—on the assumption that certain emotions are ways of apprehending value (Pizarro, 2000; Little, 1995; Zagzebski, 2004). Properties, functions, and the capacity to take an interest in x, are all abstract and impersonal, Ladmaker is not. Narratives and reflection on emotive states bequeath a knowledge of persons as worthy.

7.2.2 Intention: Identifying the Human Action

It is important to be clear that, by intention, I mean one's reason for acting (Shaw, 2006, 2015), which includes one's beliefs about the goodness of a certain state of affairs and the means to achieve it. But intention also specifies that towards which one's will is oriented. I may have several reasons for acting in any number of ways. But doing *this* action on *this* occasion involves a reason for doing so and my "will is set on achieving it" (Pilsner, 2006, 12). Intention includes both an intellectual and volitional component. As part of the intellectual component, it includes both one's *chosen* means and end.[7] Intention is one's plan of action that requires beliefs about cause-effect relations (i.e., that this means brings about this end) and what states of affairs are good (one sets out to obtain an end one apprehends as good). At the same time, intentions should be distinguished from cause-effect relations themselves, from what an agent may know, and from the agent's desires and motivations (Pellegrino, 1996). Lastly, chosen means should be distinguished from accepted or tolerated side effects. I explain such distinctions presently.

If I pound a nail I am bringing about two states of affairs: making a noise and securing two objects. I cause both, but only intend the latter. I may know that both effects will occur, but I intend only one of them. The next two examples illustrate how desire and intention are distinguished. If I have a social phobia, I may not desire to be around crowds of people, but I would intend to do so as part of what I believe is beneficial psychotherapy. I may desire to watch lowbrow television pro-gramming in the evening, but instead I intend to read and study for an upcoming exam. The action I do in cases of conflicting desires must be intended.[8] With regard to the distinction between intention and motivation (Masek, 2009), one's actions may be motivated out of fear, altruism, and hatred, etc., but the actions partly informed by such motivations may include lying, stealing, and murder. Intentions partly specify or define *actions*; fear, hatred, and altruism are species of *motivations*. Hatred, for example, is not an action, but murder is. Murder is an act of intending someone's death to deprive the victim of a good (i.e., life). Explaining why I have the intentions I have may involve appeal to one's motivations. But the explanans and explanandum cannot be conflated.

Lastly, one's chosen means cannot overlap with unintended or merely tolerated side effects. Suppose I stand to earn a lot of money from my uncle Charlie's life insurance policy (Tollefsen, 2006). In one scenario I choose to kill him in order to obtain that money; in another scenario I do not intend his death but he dies of natural causes and I inherit the money. In the first case his death is a means to my intended end, but in the second it is clearly not a chosen means. (Strictly speaking, it is not a side effect either since it is not an effect of my actions.) Suppose I have cancer and ingest chemotherapy which causes both the destruction of cancer cells and hyper-emesis. Hyperemesis is a side effect of my action plan even if it occurs before my healing. Side effects are those effects of my actions that occur outside of what I specifically intend. The spatial metaphor of "outside" is meant to capture the idea of

---

[7] For further discussion on these opening points, see Michael Bratman (1981, 2000), Robert Audi (2001, Part II), Elizabeth Anscombe (2000), Cavanaugh (2006), Pilsner (2006), and an under-appreciated though informative work D'Arcy (1963).

[8] There might be cases of knowingly performing action A but without intent to do A. Consider cases involving duress where it is clear the agent does not desire to do X, but does X under threat. My own intuitions are hazy on whether acting under duress to do X must involve intending X. Fortunately, this complication does not affect the doctor's action in euthanasia (except in cases where a conscientious objection is overridden).

aiming at states of affairs I apprehend as good. Side effects might be accepted, tolerated, or downright repudiated, but in neither of these options do I intend a side effect. This is not to say that I cannot be held morally responsible for any evil side effects of my actions. A scientific researcher doing a phase I first-in-human trial might intend to find a cure for a rare disease and choose appropriate means for doing so, but because of negligence in reviewing preclinical evidence she can be held morally responsible for serious adverse events that were not intended.[9]

Even so, there are very good reasons to believe that intentions figure prominently in a moral analysis. J.L.A. Garcia (1997, 171 ff.) observes that the applicability of terms in our moral vocabulary such as 'lie', 'rape', or 'kidnap', require that the agent have certain intentions such as to deceive, to coerce, etc. Intentions partly fix what moral action-types there are. More importantly, intentions matter morally because they are a "form of morally significant favoring, a form of response to something—such as life or death—that has positive or negative value" (Garcia, 1997, 174). Similarly, Lynn A. Jansen (2010) comments that since intentions are one's reasons for acting, they infuse one's actions with reason. The intentions of an agent "condition the meaning of his action…The reason that guides his action conditions the meaning of what he does…And the meaning of his action, both to himself and to others, is an ethically significant factor in assessing his conduct (Jansen, 2010, 28).

These clarifications serve to correct some misunderstandings. In an uncharacteristically oblique moment, McMahan argues that even if the person has intrinsic worth, that doesn't give us an argument against preventing that person from suffering either temporarily or permanently. "If it is compatible with respect for one's worth as a person to impair one's rational capacities temporarily for the sake of relief from suffering [e.g., taking an analgesic or accepting total anesthesia] this should also be true of eliminating one's rational capacities through death" (McMahan, 2002, 481). If there is no violation of one's worth in the former case, a case which involves suppressing one's rational potencies for the sake of avoiding suffering, it should not count as a violation of one's worth if the avoiding of suffering is permanent—as would be the case if one were eliminated through death. The point for McMahan is to show how what is good *for* the person can constitute a reason for eliminating the good *of* the person. One is not violating or acting against the good of the person by giving anesthesia, so too for killing the patient who is suffering intractably.

What makes the argument oblique is that it fails to specify precisely how the person is "permanently" eliminated. McMahan speaks of eliminating the person "through death." But this misunderstands that what is objected to in the argument for ~(pk) is a human *action* that has as its aim the elimination of a thing with intrinsic worth, not the *state of affair* of being dead.

McMahan is right that the *act* of impairing one's experience of suffering through palliative sedation is permissible; he would likewise agree that impairing someone's

---

[9] See Wilson (2009) for commentary on the famous Gelsinger case.

consciousness just for fun is not. What this reveals is that one needs a reason to suppress consciousness, and one should proceed to do so in a parsimonious manner. Just because suppressing consciousness temporarily is permissible, it doesn't follow that killing the patient is permissible. Again, what blocks such an inference is the principle of parsimony.

Clarifying what is meant by intention, means, and side effects and how these concepts function to define a human action is also necessary to rebut a similar misunderstanding illustrated by the following study. A survey of Dutch physicians involving questionnaires asked, among other things, about the reasons why the physician hastened the death of their patients by euthanasia or physician-assisted suicide. The top two most frequently cited reasons were the alleviation of symptoms via high dose opioids and the withholding of treatment. Neither action necessarily involves an intent to kill.[10]

To summarize, the argument for ~(pk) is that all human persons have intrinsic dignity, by which is meant that there is an aspect of their worth that doesn't wax or wane relative to their degree of suffering or functionality. We have a worth insofar as we exist. This worth can be "seen" through various emotive states such as regret, grief, and love. It is immoral to attack, destroy, or eliminate something that has inherent worth. Euthanasia intentionally eliminates human persons. Therefore, euthanasia is immoral.

7.3 The Nature of Love and Caring

In this section the parallax shifts from focusing on whether *the patient* has intrinsic dignity, to focusing on what is a loving and caring response *by the clinician* to those patients who are sick and suffering. The present focus is on the virtues of the clinician, not necessarily the moral status of the patient.

When a patient is admitted to the intensive care unit, the patient typically receives treatment for the disease that caused her to decompensate. In a similar way that cancer cells cause the destruction of healthy cells, and treatments aim to slow or destroy cancer cells, so too, a pathology of sorts affects the terminally ill in terms of their own moral worth. What happens is that such patients tend to lose sight of their own worth. Chochinov observes the following:

> Our own research demonstrates that patients approaching death may feel a burden to others; that life is futile, and an affliction to those they feel encumbered by having to look after them. Self-perceived burden is contagious and self-perpetuating; patients who experience it may cause family members to feel helpless and exhausted, tacitly affirming that they are indeed a burden (Chochinov, 2023, 2884).

---

[10]This is not the place to go into such distinctions as they are canvassed elsewhere, and the distinction between intention and foresee is, at least conceptually speaking, quite clear. See Stuchlik (2021), Keown (2018), Tollefsen (2006), and Cavanaugh (2006).

Chochinov is directing our attention to the fact that often concurrent with a patient's decompensation during a terminal illness runs a self-deprecating view of his or her own worth. Though this is understandable, it is understandable in a way that we understand that cancer cells kill healthy cells. We shouldn't stand back and let them do so, but we understand how such cells act. One has reason for opposing the force that illness has on one's self-perception. Chochinov calls this dignity therapy.

Dignity therapy is analogous to intensive care. Anyone working in emergency medicine or intensive care might be familiar with the ABCs of that care: airway, breathing, and circulation. Therapy for a terminally ill patients' existential and psychological decompensation also is governed by an ABCD of caring: attitude, behavior, compassion, and dialogue (Chochinov et al., 2002). Intensive care is so named because it intensely focuses on managing the patients' physical symptoms. Chochinov finds a parallel in what he calls intensive caring according to which it is a focused management of the patient's existential and psychological needs. In this section, I address two key questions: what does intensive caring look like? Included in the answer to this question is a discussion about what principal actions constitute intensive caring, and what actions are precluded by it. The second question is: Does intensive caring comport with a philosophically respectable view of what love is? The answers to these questions will form the second argument for ~(pk).

7.3.1   Intensive Caring and Dignity Therapy

A woman accompanied her husband to the emergency department with a chief complaint of dyspnea and abdominal pain. An initial workup revealed lung cancer with metastases to the abdomen. The cancer was quite advanced and he was put on comfort care. He survived the night but was somnolent the next morning. The wife of the patient relates the following exchange she heard while bedside with her husband:

> I heard someone say, "bed 1 is Mrs. X in bed 2 is Mr.----." I was focused on my husband and did not hear what was said next about Mrs. X. I then clearly heard a male voice say "Mr.----" and a second male voice say, "well, I guess we could go in and see if he's still alive." Another male voice laughed and then the group moved down the hallway. I cannot tell you how painful this was to me (Chochinov et al., 2002, 1).

In high mortality units, clinicians might get accustomed to death and lose sight of the preciousness with which such patients are viewed by their loved ones. Dignity therapy requires of the clinician to see this preciousness at all points in the patient's disease process, and aims to have the patients themselves see this preciousness at all points in their own disease advancement.

Dignity therapy can be thought of as a constellation of therapeutic modalities that serve to restore a patient's sense of his or her worth.[11] Broadly understood,

---

[11] The literature understands dignity therapy to be a specific therapeutic modality, characterized by the development of a generativity document (Chochinov, 2022). In reviewing the literature, how-

dignity therapy has as its goal the restoration of a patient's own self perception of her worth, and the means by which the goal is achieved is by way of the ABCD's noted above.

Attitude is how we see or perceive our patients vis-à-vis their worth. Since all we see or perceive the world is filtered through our expectations, beliefs, and possibly, prejudices and biases, practitioners of dignity therapy are prompted to reflect on their own way of looking at their patients. Chochinov relates a story about when his sister who had cerebral palsy and at age 55, experienced severe complications including acute respiratory distress and bodily contractures. External respiratory aids were not slowing her deterioration prompting a decision on whether to intubate her. The attending physician, relates Chochinov, "approached me with a cryptic and chilling question: "does she read magazines?" (Chochinov et al., 2002, 36). This sounds like a completely irrelevant question from an outsider, but the attending was wanting to know whether it was worth it. Does she still engage in worthwhile activities? Chochinov, being a clinician himself, new the subtext of such a question. He says,

> ...the subtext read as follows: "she looks and is twisted like a pretzel and you know we don't intubate pretzels." Realizing the gravity of his question and the implications I knew would be attached to my response hit me with a wave of nausea... I responded, "yes, she reads magazines, but only when she is between novels." (Chochinov et al., 2002, 37).

Chochiniov's point with his response "only between novels" was to rebut the prejudicial evaluation of his sister's quality of life. He comments, "there is no doubt that they saw Ellen [his sister] in a particular way, based on their own life experience and having absorbed a lifetime of messages, biases, and assumptions regarding what it means to be disabled" (Chochinov et al., 2002, 37).

Consider evidence that clinicians evaluate the quality of life of disabled patients as lower than the patients themselves evaluate it. Locked-in syndrome (LIS) is a syndrome in which the patient is fully conscious but fully paralyzed and unable to speak. It is caused by a stroke to an area of the brain stem, usually the pons. What is it like to be fully conscious and fully paralyzed? Apparently, it is not as bad as one might think. Steven Laureys, Perrin, et al. (2005) summarizes much of the literature on this topic considering quality of life measures, end of life decision making, and suicidal ideation. In all three areas, the conclusions may strike some as surprising. Ghorbel (2002; see also Doble et al., 2003) administered self-reports on mental and physical well-being. Self-scored perception of mental and physical health were not significantly different than age-matched normal controls. Leon-Carrion et al. (2002) discovered that, of 44 subjects, 48% regarded their mood as good whereas only 5% regarded it as bad. 13% noted that they were depressed, 73% enjoyed going out and 81% met with friends at least two times a month. Of

---

ever, the casual observer will find that there are several other therapeutic modalities such as meaning centered therapy, and life review, all of which aim at the same goal, namely, the restoration of a person's sense of worth. Since it is no part of my argument to adjudicate which modality is more effective, I discuss them under the heading of dignity therapy, broadly understood.

note, suicidal ideation was correlated with perception of pain, indicating that proper pain management might decrease the incidence of depression generally and suicidal ideation specifically. A subset of patients in the Ghorbel (2002) study suffered total LIS, but very few had suicidal ideation even after being locked-in for 6 years. On a scale from 0 to 10 (never to constantly), only four patients experienced suicidal ideation more than three on the scale, compared to eight who *never* had suicidal ideation and only four who rarely had it. Regarding treatment choice, 80% wished to receive antibiotics if they were to contract pneumonia and 62% elected to be full codes.

Bruno et al. (2011) explored quality of life (QoL) assessments with 91 LIS patients. Forty-seven patients expressed happiness, with only 18 expressing unhappiness. Unhappiness was associated with lack of integration into social life, anarthrea, and lack of recreational activities. Bruno et al. conclude:

> Our data stress the need for extra palliative efforts directed at mobility and recreational activities in LIS and the importance of anxiolytic therapy. *Recently affected LIS patients who wish to die should be assured that there is a high chance they will regain a happy meaningful life* (Bruno et al., 2011, 1, emphasis added).

Lule et al. (2009) discuss the "disability paradox" that this evidence suggests. Severely disabled patients, like those in LIS and amyotrophic lateral sclerosis (ALS), adapt to their disabilities as indicated in higher QoL scores the longer they have experienced the disability. Important for my purposes is the following observation, "Preliminary results from our study on clinicians' perception of LIS show that in 97 interviewed health-care workers the majority (66%) considered that 'being LIS is worse than being in a vegetative or minimally conscious state'" (Lule et al., 2009, 347). This assessment is clearly incongruent with what LIS patients themselves think.

Finally, research involving ALS patients, a motor degenerative disease leading to progressive paralysis, shows the same mismatch between clinicians' and patients' QoL judgment. A common sequelae of ALS is ventilator dependence. On the Life Satisfaction Index (LSI) (using a Likert scale 1–7), ventilator users reported a mean score of 4.98. For perspective, normal controls reported a mean LSI of 5.33. The remarkable finding in this study was not only that ventilator users had near equivalent LSI scores as normal controls but also that health care professionals' assessment *of the ventilator users'* life satisfaction was 2.42—far below the patients' own assessment of their life satisfaction (Bach, 2003, S25).

Here are cases in which the patient's self-perceptions deviate from their own clinicians' perceptions. But the numbers canvassed above are not 100%. Not all patients view their lives favorably. Are the clinicians who also view their lives as unfavorably off the hook?

Many patients look to their clinician to give them their sense of meaning and worth. It is true and understandable that many patients facing terminal illness can grow increasingly frustrated with inadequate pain control, the continuous loss of one's functional abilities, and loss of control and independence. Conversely, given "appropriate palliation and the rallying of a community of support, thoughts about the wish to die can dramatically recede" (Chochinov, 2004, 1336–1337). How so?

Nancy Lebrefe relates a story to the effect that a patient was visited by eight different people over his hospital stay, seven of whom were healthcare providers or consultants. Yet the only one who introduced herself and asked him his name and inquired into how he was doing was the food service person who dropped off his lunch. When who you are is passed over, pretermitted to use a theological term, in place of what you have, the transition from personhood to patienthood can be an assault on the self. In an incipient description of dignity therapy, Chochinov explains that it focuses on appreciating and acknowledging the personhood of the patient, thereby enabling the patient to catch a glimpse of his or her own dignity. When terminally ill patients themselves are asked what is most important to them in maintaining their own sense of dignity (the term dignity was not defined for them), appearance was rated as the most important feature for maintaining their sense of dignity. And this concern about appearance extends to how they are viewed by others. To quote at length, Chochinov explains,

> By listening to patients, our perception of who they are extends beyond the confines of their illness, thereby shifting the patient's perception of how they are seen and heard. Validation of their concerns and ascribing meaning to their experience, according to the Dignity Model, can bolster hope, even for those whose illness has long since extended beyond the reach of cure. The reflection that patients see of themselves in the eye of the care provider must ultimately affirm their sense of dignity (Chochinov, 2004, 1339).

So, clinicians can function as mirrors for their patients' own sense of dignity and worth. How the clinician views the patient can reflect back to the patient just how important the patient's existence is.

Since attitude concerns perceptions and perceptions are a function of one's expectations, background beliefs, or biases, this excursion into the empirical literature on quality-of-life assessment—both how such assessments can diverge between patient and clinician, and how the patient is vulnerable to the clinicians perceptions—is essential. The other features of dignity therapy need not detain us for long.

B is for behavior, and the principle behavior concerns what the clinician is to do during dignity therapy. There are two related goals here, non-abandonment and affirmation. "A foundational element of this approach is nonabandonment, which demands committed, ongoing care, and caring, even when patients no longer care about themselves" (Chochinov, 2023, 2884). Cecily Saunders, the harbinger for the modern hospice movement, observes that suffering is only intolerable when no one cares that one suffers (quoted in Chochinov, 2023, 2882).

Affirmation is achieved in several ways, but an essential element of dignity therapy is the drafting of a generativity document. Chochinov explains, "As a developmental task, generativity focuses on investing in those who will outlive us. Attending to generativity means finding ways of prolonging one's influence across time in the service of others. For people who are terminally ill, this means extending aspects of self up to and beyond death itself" (Chochinov, 2012, 37).

It might be hard to see why development of a generativity document has the empirical success that it does (Chochinov et al., 2005).[12] Hopelessness, for the terminally ill, is more highly correlated with requests for euthanasia or suicidal ideation than even depression. And hopelessness is defined in this context not with reference to a favorable prognosis, such patients are by definition terminal and do not have a favorable prognosis. "Rather than being predicated on time, hope, toward the end of life, is intimately connected to notions of meaning and purpose" (Chochinov, 2012, 18). The antidote to hopelessness, therefore, is to reclaim a sense of meaning and purpose. And this can be done by shifting the patient's view of one's purpose away from narrow definitions concerning prognosis to broader definitions of how that person's life contributes to and benefits others.

But this understands meaning still in exclusively functional terms. The idea is that the patient has a sense of contributing to or bringing about a good state of affairs. So it is understandable why legacy enhancing therapy is successful, namely, it allows the patient to morph her hope from desiring a good prognosis for *oneself* to a desire that *others* benefit. But there is a more radical permutation of hope that is available to patients that departs from the focus on productivity.

Viktor Frankl remarks that life can be made meaningful in at least three different ways.

> First, through what we *give* to life (in terms of our creative works); second, by what we *take* from the world (in terms of our experiencing values, be it in nature, or in culture); and third, through *the stand we take* toward a fate we no longer can change (in incurable disease, and in operable cancer or the like) (Frankl, 2010, 118).

It is this third way of acquiring meaning that is a more radical re-understanding of hope in the setting of terminal illness. It involves getting the patient to appreciate his or her worth *during* suffering. Focusing attention not on fixing suffering, but on standing with the sufferer and affirming his or her worth involves humility. The focus on fixing in the setting of terminal and incurable illness can exacerbate the patient's own feeling of hopelessness. Therapeutic humility, Chochinov writes, aims "to understand the nature of the patient's suffering, while creating a safe space to bear witness, to validate, and to comfort always" (2023, 2886). He relates a story of a successful academic whose career was cut short by complications following cancer surgery. This led to years of what seemed to be intractable depression. "When she felt close to the edge, I would remind her that our work together *mattered, that she mattered*, and that I remained committed to seeing her (Chochinov, 2023, 2886, emphasis original). After trying yet another therapeutic approach the patient observed the color of the door. Chochinov noted that the door had always been that color, to which she replied that she now cares; she cared about her environment. The therapeutic approach that best realizes Frankl's third source of meaning, particularly in the setting of terminal illness, is that "Continued involvement offers the

---

[12] There are other related modalities as well that show success. See Breitbart et al. (2000), Breitbart et al. (2018), Salamanca-Balen et al. (2021), and Vuksanovic et al. (2017).

opportunity to sustain patients and sometimes, even the potential for healing" (Chochinov, 2023, 2886).

C is for compassion which has both a cognitive and volitional component. The cognitive component is an awareness of or knowledge about a patient's suffering, and the volitional component is the will or motivation to assuage it. To assuage it does not mean that the clinician focuses on eliminating it, managing it, or otherwise trying to fix it. Some of a patient's suffering, such as the loss of one's functional abilities is intractable. Rather, compassion aims to accompany, accept, and acknowledge the patient's feelings. These feelings include grief over the loss of one's functional abilities, regret, or the depression of coming to the close of one's life. Awareness is achieved, of course, by asking the patient how she or he might be feeling, what can be done for them, and reflecting on the clinician's own actions in regard to how the clinician might make the patient feel. The D in dignity therapy is self-explanatory.

The point in this discussion is to limn a way of addressing a patient's existential suffering which forms the basis for most requests for euthanasia. The core idea with dignity therapy is to restore the patient's own sense of her personhood—understood in this context as the perception of one's worth in spite of one's suffering. The means to this therapeutic goal is to imbue the patient with a sense of meaning and purpose, broadly understood as discussed above. The attitude of the clinician should always be animated by a sense of the patient's worth. Instead of asking "does she read magazines?" dignity therapy asks "how can I help this patient see her worth?"

Dignity therapy, understood to include several different therapeutic modalities, has been clinically tested and proven to be quite successful in restoring a patient's hope, meaning, and general life satisfaction. The successes here are not ubiquitous, but much better than standard approaches. What about those who don't respond to dignity therapy? May not euthanasia be permissible for them?

Such questions would miss the point of the extended description of dignity therapy provided here for two reasons. First, the point of that therapy is for the clinician to restore a patient's view of the patient's own worth and never to abandon this goal of restoration. Eliminating the patient through euthanasia would so abandon the patient. The second reason is that the description captures what loving patients look like who face terminal and incurable illness. Loving them requires standing with them while they are suffering. To argue for this second reason, however, requires a brief description of love.

Regarding the notion of love, there are two views that I canvas here. There are of course other views, the ones I choose are relevant for the discussion on euthanasia as will be seen.

7.3.2   An Account of Love

The first view of love has it that love is a response to what the lover perceives to be valuable features about the beloved. Love is engendered by what the lover is attracted to in the beloved. The reason for loving the beloved is because the beloved has valuable features, properties, or characteristics. Eleonore Stump refers to this as the responsive account of love (2010). The responsive account has considerable psychological appeal particularly if we ask the lover why she loves the beloved. In ordinary conversation, what follows is a list of properties, features or characteristics.

One can also see that the responsive account captures best the view of love implicit in the neutral container view of life's worth (discussed in the previous chapter). On that account, you and I are valuable if we have valuable features, experiences, or abilities. So, if the responsive account fails to capture our intuitions on what love requires, the proponent of (pk) would be hard-pressed to find a replacement.

On the responsive account, love is a response to appealing characteristics. Since other persons do or could have such appealing characteristics, Stump observes, "another person could be acceptably substituted for the beloved, provided only that the new person had a valued characteristic of the beloved" (Stump, 2010, 86). The responsive account gets the wrong result here. Intuitively, love is nontransferable in the sense that the beloved cannot be substituted with another, even if the other possesses similar characteristics. We have already encountered this feature of love in Brewer's and Gaita's accounts explained above. Quoting Robert Kraut, Stump affirms, "the non-transferability…of love is a defining condition of its being directed toward a unique individual" (Stump, 2010, 86).

A second problem with the responsive account is that since love is a response to a person's characteristics or properties which can be had in varying degrees, it seems to follow that the strength of one's love should vary in proportion. This would render love fragile and inconstant. Harry Frankfurt observes that "it is quite clear to me that I do not love [my children] more than other children because I believe that they are better [than other children]."

Related to the previous problem is that we expect love to endure and be resilient through various changes. When valuable characteristics either lessen or cease altogether, the responsive account has it that love should also lessen or cease. This idea of resilience, a corollary to fragility, should be a feature of one's understanding of love.

Though the focus in this section is on what a loving clinician looks like, it is important to observe here that these features apply to the patient's own self charity. If a lover ceases to love the beloved because the beloved could not write quality novels anymore, or who lost one or more of one's senses, or who became mild or moderately incontinent, would be a shallow, fragile, and callous lover. We should see, therefore, that when a patient fails to love herself for any of such reasons, rather than confirming her in such misperceptions, we should treat them, and help such patients become more resilient lovers of themselves. Summarizing the problems with the responsive account, Stump has readers take on the position of the beloved who asks "why do you love me?" For the responsive account the lover must respond, "because I love in you characteristics X, Y, and Z, which I find intrinsically valuable" (Stump, 2010, 87). Clearly, *you* have dropped out as the proper subject of the lover's love.

The responsive account seems to be the only account of love that fits the ontology of Rachels, Kohl, and the other proponents of the neutral container view. Insofar as the responsive account fails to capture our intuitions on love, it looks like a defense of (pk) would be incompatible with a satisfactory account of a clinician's love.

A better account of love is to understand it as willing the good of the other. Understood as willing the good of the other, such an account does not violate the non-substitutability of love. I can will the good of *this* person, not this or that feature, characteristic or functional ability. Furthermore, love understood as willing the good of the beloved admits of resilience and non-fragility since the lover can will the good of the beloved whether or not or to whatever extent the beloved realizes such and such goods.

There is, of course, much to say about the notion of love as such. Here it is enough to see that this account of love, which satisfies the conditions of non-substitutability, resilience, and non-fragility, is what dignity therapy prescribes for patients experiencing existential suffering. In so far as dignity therapy focuses on restoration of *this* person's own sense of worth, it is nontransferable. It is resilient in so far as it walks with patients to their natural death. And it is non-fragile in the sense that dignity therapy does not change its goal of care, so to speak, when the patient is suffering. The goal of care is always to *restore* this patient's own sense of worth as much as this patient can at this point in the disease process.

7.4 De Jure Defects

In Sect. 7.3, I considered two arguments for ~(pk). I will refer to them as the dignity argument and the argument from love respectively. The dignity argument runs roughly as follows:

(1) All human persons have intrinsic dignity, (by which is meant that there is an aspect of their worth that doesn't wax or wane relative to their degree of suffering or functionality).
(2) It is immoral to attack, destroy, or eliminate something that has inherent worth.
(3) Euthanasia intentionally destroys or eliminates human persons.
(4) Therefore, euthanasia is immoral.

The key premise in this argument is, of course, premise (1). The argument for it considers different vectors by which we can see or appreciate a person's intrinsic worth. The argument from love runs roughly as follows:

(5) Loving another must be nontransferable (in the sense explained above), resilient, and non-fragile.
(6) If dignity therapy aims to affirm and restore this patient's own sense of worth even in suffering, it loves another in a way that is nontransferable, resilient, and nonfragile.

(7) Dignity therapy aims to affirm and restore this patient's own sense of worth even in suffering.

(8) Therefore, dignity therapy loves this patient. But it is also true that if one loves this patient, one would engage in something like dignity therapy. One would aim to restore the patient's sense of his or her worth.

(9) Dignity therapy is incongruent with euthanasia (one cannot aim to restore a patient's sense of worth by eliminating the patient).

(10) Therefore, euthanasia is incongruent with loving the patient.[13]

Understood as such, the de jure defects come to two, one for each argument. For the dignity argument, the defect is that it relies on the reliability of our emotion-value perception, for example, our responses to the Charles-Ladmaker narrative. The problem might be best described autobiographically. When I first read the story, I immediately saw the implications for euthanasia. The kind of character I saw in Charles is someone who walks with and tries to restore Ladmaker to as much functionality and comfort as he could provide for him at that time. Putting Ladmaker out of his misery by intentionally killing him was clearly not a proposal he considered seriously. At great cost to his own health, Charles loved Ladmaker. The precise feature that made Charles's love so laudable was its resilient aim to restore. When I let students read this narrative, however, many did not make the connection with euthanasia at all, even after asking them if there is any connection with euthanasia. They saw that Charles' choice of action was one among many that might be *permissible* in that scenario. They might admire Charles, and at the same time tolerate that if he had killed Ladmaker, that would have been permissible.

The de jure defect is this: through grief, remorse, and love we might catch a glimpse of a person's inherent worth and what might be laudable or *supererogatory* about caring for the person, but that might be compatible with euthanasia being *permissible*.

This is a very weak objection, but one could make it if one presumes (pk). I think it is weak for the reasons already noted in the previous chapter, namely, it must presume that persons do not have inherent worth. It is also weak in so far as it ignores what might be admirable about Charles aiming to *restore* Ladmaker with such resilience. Gaita's explanation for what makes it admirable is that Ladmaker's life is as valuable as Charles' love showed it to be.

The de jure defect for the argument from love is a permutation of the same point. Nowhere in the literature that describes dignity therapy is it explicitly stated that it is incompatible with euthanasia. Indeed, the goal of care for dignity therapy is to restore the patient's own sense of his or her worth. But if this restoration fails, is it not *compatible* with dignity therapy to euthanize those who are recalcitrant to its therapy? Granted this would eliminate the suffering by eliminating the sufferer. And if the dignity argument is correct, doing that would be impermissible. But what does the argument from love have to say about the lingering possibility of dignity therapy

---

[13] Symbolizing the argument from (8)–(10) might run as follows. (8) $D \equiv L$. (9) $D \rightarrow \sim E$. Assume L. It follows that D from (8). From (9), it follows that $\sim E$ by *modus ponens*. Therefore, $L \rightarrow \sim E$.

failing? Consider an analogy. Suppose that chemotherapy fails to stave off a patient's cancer. Euthanasia is not *incompatible* with *chemotherapy* in that scenario because the therapy failed, it no longer applies. It looks like any goal of care would be possible in the setting of such failure. Conceptually speaking, in the setting of dignity therapy's failure, intentionally ending the patient's life is not incompatible with it.

This defect too is weak, but one could make it with the right presumptions in place. The reason I think it is weak is because the orientation of the clinician's mind towards the patient must be focused on the patient's restoration. Recall the story of the academic patient whose career disintegrated after complications following cancer treatment. Her existential suffering was in the main, intractable. But Chochinov's therapeutic focus was *always* on making her feel that *she* mattered. So, dignity therapy is incompatible with intending to kill one's patient insofar as it adopts a resilient stance to restore the patient's sense of his or her worth. And there is a sense in which it never fails if we focus on the goal of restoration understood relatively, i.e., as restoring the patient to the highest functionality the patient is capable of at that point in the disease/illness process (Thomas et al., 2024).

These two defects share a common feature, namely, they are defects having to do with certain moral intuitions one's interlocutor might have. We might tentatively identify these as justification-defects satisfying condition (iv). But their status as being tenuous qua defects is because of this very feature. Having a different appraisal of Charles's actions or Ladmaker's worth is to be expected at this more fundamental level of moral discourse. Either you see it or you don't. And if you don't, it doesn't follow that arguments based on the value commitments are defective arguments. The *argument* that justification J is implausible (condition (iv)) is absent. That a premise is subject to peer disagreement exerts epistemic effects both ways, whether we affirm it or deny it. For example, in the previous Sects. 6.1 (i.e., discussion of Lachs) and 6.2.5. I discussed the appeal to one's intuitions on killing in certain cases. I noted that what we wanted was an argument for (1), not to assume it. That same type of defect recapitulates in the present chapter in the discussion concerning dignity.

In any case, the defects we don't find in the arguments from dignity and love are defects that are more structural in nature. Defects such as circularity which affects the TRIA justification for abortion, or non sequiturs like the problem of underdetermination which affects the no person argument for abortion, or the structural defects affecting the arguments for death being a benefit (i.e., Sect. 6.2.3), or that consent gives one reason to kill the consenter (Sect. 6.2.2).

So, each argument falls prey to different kinds of defects, with differing gradations of severity. Defects such as circularity are quite severe, defects such as peer disagreement about the moral valence of a case are less severe. The argument for ~(pk) falls prey to the latter, the argument for (pk) (for both abortion and euthanasia) falls prey to both. The next and final chapter outlines what practical implications follow from this read on the quality of such arguments.

Chapter 8
Moral Risk

8.1 Introduction

The plan for the present chapter is to return to the argument presented in the intro-duction and to complete its defense. The overall argument of the book is as follows:

(A) The arguments for (pk) suffer from serious de jure defects, and the arguments for ~(pk) suffer from trivial or much less serious de jure defects.
(B) If the moral risk in being wrong in acting on (pk) is greater than the risk of being wrong in acting as if ~(pk) and the justification for (pk) suffers from greater de jure defects in relation to the justification for ~(pk), acting as if (pk) should not be done
(C) The moral risk in being wrong in acting on (pk) is greater than the risk in being wrong in acting as if ~(pk).
(D) Therefore, acting as if (pk) should not be done.

The previous chapters defended premise (A) substituting for (pk) abortion and euthanasia. What remains to be defended is the *inference* in premise (B), and prem-ise (C). These are defended in Sects. 8.2 and 8.3 respectively.

8.2 Moral Risk and Practical Rationality

A brief review of Chap. 2 and the roles of moral risk may be a starting point for the present discussion.

1. Risk might be considered relevant in adjudicating who in a dialectical exchange bears the burden of proof. The basic idea is that if risks are not symmetrical, the proponent of the riskier position—the position that if one is wrong involves greater harms—bears the initial burden of proof.

S. Napier, *Justified Killing*, https://doi.org/10.1007/978-3-032-14946-6_8

2. A risk in being wrong about P might make P more sensitive to defeaters. Quite independent of whether or not the proponent of P bears the burden of proof, one could say that a higher risk in being wrong about P renders P more easily undermined by counter considerations. This is the defeater role.
3. A risk in being wrong about P renders one's action on the basis of P immoral (or insufficiently justified). On this understanding, the relevance of risk has less to do with the *epistemic standing* of my belief that P, and more to do with my *moral responsibility* to consider the potential harms to others in being wrong about P. Chap. 2 discussed two different positions. Briefly, one might think that a belief with higher stakes impugns my claim to know that claim (Fantyl and McGrath, 2009); or the moral risk allows that I know the claim, but it would render me irresponsible in acting on it (Reed, 2012). This was called the bar of justification role—or simply the justification role.

Each of these roles pertain in some way to the actual justification used to support a claim. With these roles in mind, we can motivate the inference in (B).

There are numerous cases that can illustrate the effects of moral risk. Most of the literature on what is referred to as pragmatic encroachment focus on what is referred to as the bank case (DeRose, 1992, pp. 913 ff.).[1] In case A, Kermit and Piggy are driving home on a busy Friday evening. Kermit has just been paid and he is considering depositing the check in the bank that night. But as they drive past the bank they notice that the lines are exceptionally long. Kermit proposes waiting until Saturday morning to deposit the check having remembered that the bank was open on Saturday several weeks ago. Piggy mentions that many banks are closed on Saturdays. Kermit continues to believe (suppose rightly) that the bank will be open on Saturday relying on his memory of the moderately-distant past. Case B is just like case A except that Kermit and Piggy have a very important mortgage payment that will be withdrawn automatically over the weekend and their account is insufficient without the deposit. If they do not deposit the paycheck, they will be evicted from their home. Kermit recalls that the bank was open on Saturday several weeks ago, and Piggy mentions her potential defeater to that belief. Does Kermit know that the bank will be open on Saturday in case B? Many authors have the intuition that Kermit does not know in case B, but he does in case A. The explanation is that the cost in being wrong that the bank will be open on Saturday in case B is too much to rely simply on memory of the moderately-distant past. Furthermore, banks could change their hours, it could have been opened on a Saturday just once a month, etc.

Cases involving practical costs can be interpreted in at least three different ways. First, one could say that Kermit does not know that the bank is open in case B, but he does know in case A. Second, he does not know in either case. Third, he does know in both cases but, for case B, he lacks sufficient justification to act on that

---

[1] Much has been written on pragmatic encroachment. For book length treatments, consult Fantl and McGrath (2009), and Stanley (2005). Important article length work not discussed here is Guerrero (2007). Guerrero and I share numerous intuitions; my disagreement pertains almost exclusively to his discussion of abortion.

belief—the belief that it is open on Saturdays. On this third interpretation, knowledge does not require certainty, but he needs stronger justification that the bank is open on Saturday because of the costs in being wrong in case B. All three interpretations grant that, in case B, Kermit's justification—as described—is not sufficient to justify the act of waiting until Saturday to deposit the check. The strength of justification sufficient for *knowledge* comes in degrees; and one can meet that threshold before meeting the threshold of justification required to justify *acting*. In cases involving high costs in being wrong, there is not enough epistemic justification to justify acting on p, but there may be enough to justify knowing that p. This is Baron Reed's assessment of how moral risk works (Reed, (2012)), and I'm following it because it is in the main, the most ecumenical position.

Important disagreement exists as to whether knowledge covaries with the level of moral risk at stake. It appears, however, that most interlocutors to this debate agree that agents in high risk scenarios are *not justified in acting* on their beliefs; in case B, Kermit is not justified in waiting until Saturday to deposit the check. This is one reason for favoring Reed's approach. Here is another.

Reed's position avoids the problem of indifference according to which I can get knowledge simply by caring less. The problem of indifference highlights the feature of most moral risk cases according to which the agent in question cares about the values at stake. Kermit, for example, cares about having a home. We can imagine an agent not caring about that. If so, in cases where the agent in question cares less, cases A and B look epistemically equivalent. By caring less, Kermit knows that the bank is open in Case B (if we say he knows in Case A). The idea that knowledge would wax and wane according to one's cares is counter-intuitive. Though there are plausible rebuttals to this problem (Coss, 2018), avoiding this dispute with no other intuitive costs is philosophically economical. Reed's position avoids this problem by not understanding moral risk to be knowledge destroying. Rather, even if one thinks that Kermit knows in case B, most everyone agrees that he is not justified in acting on that belief given his justification. So, moral risk affects how strong one's justification needs to be in relation to one's action plan. Notice that in case B, Kermit has justification for thinking that the bank will be open on Saturdays—he has a tenuous memorial belief to that effect. What he lacks is justification *strong enough to act* as if it is open on Saturdays.

Once we see the pattern, examples abound. Consider Braddock's opening case (2024) discussed previously. A drone operator, on the basis of testimony, launches a smart weapon at the home of a putative high-ranking target. As the missile approaches the camera reveals a very low resolution image of what looks like a small child who rounds the corner of the house in time enough for the drone operator to disable the missile or veer it away from the target. The operator questions the image on the screen, and the operator's cohort who is viewing the same image confirms that it is a dog, not a small child. The operator lets the missile proceed to the target and seconds later the target is annihilated. The moral risk in being wrong that the image was that of a dog was extremely high. (Notice that the drone pilot and others involved were not uncertain about the moral impermissibility of killing innocents. This is one difference between moral risk and that of uncertaintism, discussed below).

Approaching issues of central concern in this book, Alexander Pruss (2011) reveals the role of moral risk with the following thought experiment. Suppose you accept a metaphysical view of the person according to which you and I are psychological entities who do not come into existence until after most abortion are performed. Label this view Pe (psychological entities). Now consider the following wager: "Suppose Sam is a supernatural being that you know to be incapable of asserting a falsehood… Sam tells you: "I know whether [Pe} is true or not. Here's a deal. If [Pe] is true, I will give you $50,000. If [Pe] is false, I will kill your brother. Do you accept?" (2011, p. 29). Though the case is contrived, and may be too abstract or artificial, most would not accept the wager. Pruss' point is to reveal how the moral risk in being wrong about Pe means that *a person is killed*—and few would accept that risk.

The point of these cases is to motivate the inference in premise (B) without assuming (pk). Section 8.3 argues that the moral risk between (pk) and ~ (pk) is asymmetrical. Before addressing (C), a brief discussion of ideas related to moral risk is required.

The notion of moral risk here should be disaggregated from related concerns under the broad heading of uncertainty. There are different aspects of a moral action about which one can be uncertain. One might be uncertain of the wrongfulness of an action as such, (for example, given contemporary sensitivities, the immorality of adultery or the justice of reparations). One might be uncertain not about the wrongfulness of the action, but about whether the kind of action performed in this circumstance is a species of that wrongful action. So, for example, one might be certain that human actions that involve intentionally killing human persons are wrong. But one could be uncertain whether this particular action in this circumstance counts as either an *intentional* killing, or an intentional killing of a *person*. The spelunker case (Foot, 1967) is an example where one may be uncertain that blowing up the corpulent man in the cave counts as an intentional killing.

Following Boonin (2003, pp. 310 ff.) there are different ways of managing these uncertainties but he thinks that his objections cover all cases of uncertainty. A policy of disaster-avoidance, for example, might recommend that if there is a nonzero probability that one might be wrong, disastrously wrong, one should believe otherwise. Boonin explains, "as long as we are not literally certain that an act is morally permissible, we must accept the claim that it is morally impermissible" (Boonin, 2003, p. 314). He is correct to point out that we simply do not act this way. There is a nonzero probability that ecosystems, trees, vegetation, nonhuman animals, and so on have the same right to life as you and me. This would make extricating weeds in my backyard tantamount to serial murder.

My first observation is that Boonin's objection to this way of managing uncertainty is not identical to the notion of moral risk used in the present work. If there is a moral risk in being wrong, it does not follow that one *should change one's beliefs*. I have not advocated for changing one's beliefs.

A second observation is that Boonin's strategy is to consider an uncertainty principle but change the belief content on which that principle is meant to apply. The idea is to change that content until one gets counterintuitive results. I agree with Boonin that the very low probability of my lawn, for example, having a right to life should govern my actions even if I risk mass murder in mowing it. I disagree only with the inference he draws from this procedure, not his moral intuitions.

Boonin appears to infer from this procedure that uncertainty or moral risk should not factor at all in our practical reasoning. Certainly, Boonin's intuitions on the lawn case are apt, but he should just as well sympathize with those who think that Kermit in case B should deposit the check on Friday evening. Moral risk should exert some epistemic effects on the justification for our actions. Boonin's procedure does not impugn this latter claim.

Another feature of Boonin's procedure is that he modifies the belief content to the point where it holds little plausibility—this feature is what generates the counterintuitive results. That is not the notion of moral risk employed in this present work. Rather, moral risk pertains to the risk in being wrong about (pk) *given* the typical justifications for (pk). The uncertainty arguments Boonin entertains can be explained without any reference to the actual justification for the claim in question—just as Wager arguments can be explained without any advertence to specific theistic arguments. Moral risk is different in that it applies to a claim given its standard justification. Moral risk does not increase as the magnitude of *possible* harms increase. It varies depending on the nature and quality of *justification* used to defend judgments that if acted upon have a risk in being wrong. Moral risk is dialectically contingent in that it depends on the quality of argument. Moral risk is a relational feature between the state of justification for a claim and that claim's moral consequences if it is false.[2]

One might object, however, to the focus on the relational features of the argumentative dialectic. Brian Weatherson claims that concerns about moral risk (understood as I have understood it) must think that our actions and beliefs are guided by external norms. Normative externalism is the view that our actions should be guided by the objective rightness or wrongness of the norms that guide those actions. The rightness or wrongness of the norms is given by moral reality not what the subject believes or does not believe. But normative externalists, according to Weatherson, think that,

> the most important norms concerning the guidance and evaluation of action and belief are external to the agent being guided or evaluated…. What one should do, or should believe, in a particular situation is independent of what one thinks one should do or believe (2014, p. 141).

Those who think that moral risk is an important epistemological problem believe that agents should not be guided solely by norms that they subjectively endorse. Weatherson thinks this is epistemically vicious, he calls this a moral fetish.

---

[2] Of course, in the field of human moral action, things can get complicated. See Guererro (2007) for how to manage some of these complications.

Weatherson's basic point is that concerns about risk aim to answer the question what should an agent do if an agent is uncertain about P. For Weatherson, it does not appear to matter what the reason for such uncertainty is. That is, it doesn't matter what one's justification looks like for Weatherson. The centerpiece of his argument is, rather, a distinction between two different moral motivations: desiring that one act on the rights and obligations that one sees as being the best in the circumstances in question versus the desire to act on the rights and obligations whatever they might be, or as such. In the latter, one's moral motivation is de dicto in the sense that it is a desire to instantiate what one's obligations are qua obligation. The former motivation is de se in the sense that one is motivated to instantiate goods that one apprehends at that time and in those circumstances. To be motivated de dicto is to have a moral fetish according to Weatherson. The picture we are given is of an agent who has pretty good reasons for thinking that P, but is not certain about it. If, Weatherson claims, such an agent were to suspend acting on P, for whatever reason, he would be a moral fetish.

Though Weatherson mentions uncertaintism, his argument rebuts views that think that our beliefs or actions are subject to objective norms—and in this regard his comments have something to say about my notion of moral risk.

In response, I either do not understand Weatherson or I do not think that his key claims are remotely plausible. No claim is made about what might be wrong with moral motivations de dicto. I am flummoxed on two counts. First, I don't know why Weatherson refers to the second moral motivation as de dicto since the motivation is not to obey a proposition. Rather the person's moral motivation is to realize one's obligations, and those obligations are objective features of moral reality, they are the furniture of the moral world. And obligations are only as good as the goods or values they aim to protect or promote (Wolterstorff, 2008). When we say that one obeys a command or an obligation, we are not obeying *a proposition*. Maybe what Weatherson means is that the second motivation is a motivation to instantiate whatever one's obligation is no matter the content. I don't know of any view that would endorse this understanding of moral motivation; and divine command theory, properly understood, is not such a view (see Evans, 2013).

The real problem is that Weatherson does not appear to consider the epistemic gap between our moral beliefs and moral reality. Consider his point as imported into the dialectic on killing. Both the pro-choice position (which endorses the no person strategy) and the pro-life position agree with the moral norm that one ought not to kill persons. The disagreement between them comes to whether an *actual* manifestation of consciousness is a necessary condition for being a person (call this view actualism). If it is, the no person strategy would be justifiable. I argue in Chap. 4 that the original brain transplant thought experiments do not provide any justification for actualism. What would Weatherson say? Weatherson's point seems to be that if S agrees that such a view is underdetermined and that gives one reason not to act as if actualism were true, S is suffering from a moral fetish. If that is what having a moral fetish is, then everyone should have a moral fetish.

But maybe this is not correct as an interpretation of Weatherson. So, suppose instead that S considers the arguments of Chap. 4 but still believes that actualism is justified. Weatherson's disavowal of normative externalism suggests that in this scenario S can continue to believe actualism for any reason whatsoever. It would not matter, for Weatherson, that the arguments for why the justification for (pk) suffer de jure defects are plausible or that the inferences are objectively correct. It matters only that the agent sincerely believes (pk) to be correct. If denying that this is what matters qualifies me as having a moral fetish, I gladly accept the diagnosis. Caring deeply about what one's obligations really are strikes me as an intellectual virtue, not a vice which is suggested by the sobriquet 'fetish'.

Alexander Guerrero's treatment of how knowledge interacts with moral culpability is intuitively plausible and carefully constructed. The centerpiece of Guerrero's treatment is the principle 'don't know, don't kill' (DKDK) which reads,

> If someone knows that she doesn't know whether a living organism has significant moral status or not, it is morally blameworthy for her to kill that organism or to have it killed, unless she believes that there is something of substantial moral significance compelling her to do so (Guerrero, 2007, pp. 78–79).

Two important observations are worth noting. First, the comma suggests interpreting DKDK as a disjunction, but that is incorrect for what are obvious logical reasons.[3] Instead, DKDK should be understood as a conditional as follows:

> DKDK $=_{df}$ if S knows that S does not know the moral status of an organism O *and* S does not have a morally compelling reason to kill O, it is morally blameworthy for S to kill O.

To remain consistent with the notation throughout this book, I take the phrase "it is morally blameworthy for S to kill O" as equivalent to ~(pk). And since Guerrero is discussing DKDK's application to abortion, the phrase 'not knowing the moral status of O' is equivalent to 'not knowing that abortion is permissible'.

A second observation is that what matters for Guerrero is that any type of harm is subject to DKDK. If I know that I do not know whether this thing has moral status, I ought not to *mistreat* it in any way, no matter what it is.

Third, the antecedent requires knowing that one does not know. Call the latter first-order knowledge and the former second-order knowledge. The object of one's second-order knowledge is one's own epistemic house, so to speak. The object of first-order knowledge is whether abortion is morally permissible.[4]

---

[3] A disjunction is logically equivalent to a conditional with an added negation of the antecedent. One can surmise quite clearly that denying the antecedent of DKDK does not imply that one has a morally compelling reason to kill. The conditions for having a morally compelling reason to kill something is something like self-defense against a lethal attacker.

[4] Knowing negative existentials is a bit tricky, for if there is no first-order knowledge, there is nothing there for second-order knowledge to know (Cartright, 1960; Gale, 1972). But we know that certain things do not exist. So, suppose that such standards are met when the claim pertains to one's own first-order knowledge.

What would DKDK say about abortion? And how, if at all, does DKDK differ from the notion of moral risk (i.e., premise B) discussed here? Consider the former question first.

Guerrero begins by entertaining what he refers to as an unusual case according to which the mother is both uncertain about the moral status of the fetus and she does not have anything of "substantial moral significance compelling her to get the abortion" (2007, p. 90). She simply dislikes the inconvenience of being pregnant. Assume she voluntarily engaged in sexual intercourse. Guerrero wants to say that DKDK "would imply that she is morally blameworthy for getting the abortion, given her uncertainty about the moral status of the fetus…" (2007, p. 90).

As noted, Guerrero is quick to point out that this case is unusual. The usual cases in which a mother is contemplating an abortion are ones in which she is either not uncertain about the moral status of the fetus or she has morally significant reasons to get the abortion (2007, pp. 90–91). What counts as a morally significant reason? Guerrero thinks that someone could reasonably believe that the following reasons are morally significant reasons to have an abortion:

> (a) concerns about having a child might interfere with one's life plans and happiness, (b) concerns about the financial and emotional difficulty of raising a child, (c) concerns about the actual experience of and health concerns associated with being pregnant and giving birth, (d) possible social stigma or punishment associated with having a child,… and so on (2007, p. 91).

Guerrero's point is that in the usual cases, the cases in which a mother has these concerns, DKDK does not apply. To say that DKDK does not apply means that DKDK does not entail that the woman should not procure an abortion. Conversely, she could procure an abortion and not violate DKDK.

To summarize, Guerrero concludes that the only cases in which DKDK applies are ones in which the person believes "*both* that she is uncertain about the moral status of fetuses, and that nothing of substantial moral significance compels her to get the abortion" (2007, p. 92). The usual cases, of course, are ones in which both conditions are not satisfied—one either has morally compelling reasons or she is certain of the fetus's inferior moral status. It follows that DKDK does not apply to usual cases.

DKDK does not apply if any one of the three antecedent conditions are not satisfied. In keeping with the language above, the first condition (i) is that the agent fails second-order knowledge (that she does not know that abortion is permissible). The second condition (ii) is that it is not the case that she does not know that abortion is permissible. Satisfying the second condition entails that S knows that abortion is permissible, viz.,S knows that (pk). And the third condition is if S has a morally significant reason for harming/killing O. Satisfying any one of these antecedent conditions means that the antecedent of DKDK is not met, and therefore nothing follows regarding ~(pk). I argue that on this understanding Guerrero's argument that DKDK does not apply to abortion is viciously circular or counter-intuitive. Any justification for the claim that it does not apply requires already assuming (pk).

Consider the first condition first: S does not know that she does not know the moral status of O. One can imagine a case in which someone does not know that he does not know and should most certainly not harm or kill. A general might come to the right conclusion, we may suppose, that a military intervention satisfies just war criteria. But we may also suppose that he does not care that it does. He just wants to exercise his military force against the enemy, and he can't wait for Congress to give him the clearance to do so. The general does not know that the intervention is just because he has no justification for it. And we should say that he should not act because forming judgments about killing should be done with considerable care, much more than is exercised by the General. Cases like these can easily metastasize. All that is required is that one is epistemically lazy or careless in monitoring the epistemic status of her beliefs or her inquiry. What follows is that if one fails second-order knowledge, it does not intuitively follow that one may permissibly kill. Failing the first condition represents a rather weak argument for why DKDK would not apply in the abortion case.

Not satisfying condition (ii) clearly implies (pk). DKDK requires that the agent not know that abortion is permissible. So, if she fails this condition, she knows that it is, she knows (pk). Assuming that she *knows* this requires assuming that abortion is permissible.

That leaves condition (iii). Guerrero thinks that reasons (a–d) quoted above count as morally significant reasons. Further on, he adds a fifth one pertaining to bodily rights. He states,

> And there are other arguments which conclude that even in cases in which the woman's life isn't in jeopardy, she has a right to do what she wants with what is inside of her body, and so could choose to abort the fetus even without examining the 'competing rights' claims that each might have (2007, p. 92).

In these contexts, he is assuming that it does not matter whether the agent also thinks that the fetus is a person.[5]

Consider first this right to do "with what is inside of her body." Simply having this right does not entail or give one a reason to kill the developing human being. If it were such a reason, all pregnant women would have a morally compelling reason to kill their developing babies. They may not act on it, but they have it as a

---

[5] The second condition of DKDK is originally phrased as if S does not know *the moral status of the organism O*. Guerrero seems to think that S can know that the thing in question has the moral status of a person and yet S may still have morally significant reasons for killing him or her. One could, on Guerrero's view, not satisfy the second condition, but satisfy the third. Given Guerrero's discussion of the abortion issue I specified the phrasing of the second condition according to which S does not know that abortion is permissible. On this phrasing, one could not consistently fail the second condition and yet satisfy the third. Failing the second condition means that S knows that abortion is not permissible, and if S knows this she cannot have a morally significant reason for killing anyway. As the argument in the text illustrates, the second and third conditions stand or fall together. Furthermore, I do not think that a thing's lesser moral status is itself a reason to kill it. If it is permissible to kill animals, it certainly is not *simply* because they have lesser moral status. We don't kill animals for fun, we may kill them only for food or in self-defense. Their lesser moral status is not itself a reason to kill them.

reason. But that position is absurd. Other absurdities are in the near vicinity. Does Guerrero really mean that women have a right to do what they want with what is "inside" of them? What are the contours of such a right? Do women, or any of us for that matter, have a right to give ourselves kidney cancer? Do we have a right to make ourselves throw up after we eat? Do we have a right to have our healthy limbs amputated to satisfy a sexual fetish such as apotemnophilia? Even if the reader might think that such actions are permissible, it is absurd to assert that they are permissible because we have a *right* to do them. Of course, more can be said here, but see the discussion in Chap. 3.

The real issue concerns the work that (a)–(d) are supposed to do. Consider the first reason according to which having a child might involve a change in one's life plans, or a diminution in one's happiness. The fact that one's life plans would change if one has a child doesn't at all provide a reason for killing the child. Since these are being offered as morally significant reasons, and such reasons are an independent criterion from the personhood of what is killed, it is fair game to export this reason to other contexts. If the mother's husband or boyfriend made her unhappy, does she now have a compelling reason to kill him? The fact that someone's existence would result in my being less happy, doesn't entitle me to kill that person. Reason b cites concerns about the financial difficulty of raising a child. But even here, such a reason is hardly a reason to kill one's born children. We do not solve the problem of poverty by killing the children in the impoverished area. Guerrero is usually very careful and enlightening, but these suggestions are not persuasive.

None of the reasons (a)–(d) are morally compelling reasons that would justify killing one's born child. But if one thinks that they don't function as justifications for killing one's born child, but they do for killing one's unborn child, one must already assume (pk). If the only difference is that in the latter case, the killing is an abortion, (pk) is presupposed. One must already assume that unborn human beings have lesser moral status to the extent to which it would be justified to kill them. So, the argument that DKDK does not apply to abortion in the usual cases must presuppose that abortion is permissible. If that is presupposed, then of course DKDK does not apply. But that is clearly a viciously circular argument.

Here's how I think Guerrero should be understood in connection with these previous observations. He might say that someone *could* think that bodily rights considerations are good justifications for killing. DKDK is phrased as "if someone knows that she doesn't know…" And it seems that one could mistakenly, though inculpably, think that concerns such as (a-d) are good justifications for (pk). If so, DKDK does not apply—the agent would not *know* that she doesn't know. This way of understanding Guerrero obviates the circularity in that one could fail to know that she (mistakenly) knows the moral status of the fetus or that abortion is permissible. She might do something wrong by procuring an abortion, but DKDK is not the reason why. This understanding also obviates the absurdities since there are those who mistakenly think that (a)-(d) might count as morally compelling reasons. Guerrero can agree that they are absurd, but there are those out there who don't think that they are absurd. And if so, they don't satisfy the antecedent conditions of DKDK.

Understood as such, however, DKDK turns into a very weak principle according to which someone can avoid its implications even if they unknowingly endorse absurd ideas, or endorse reasons that don't turn out to be reasons for killing at all. Such an interpretation of DKDK, however, works against Guerrero's largely correct and careful discussion of culpable ignorance. Guerrero thinks correctly that someone can be culpably ignorant—the ancient slaveholder should have known better. As argued above, thinking that reasons (a–d) function as sufficient justifications for killing one's unborn child are likely candidates for making a culpable epistemic mistake. Widespread cultural acceptance of harm or killing does not exculpate.

So, Guerrero is correct that if an agent knows that she doesn't know 'the moral status of this thing she's about to kill,' then she should not kill. Furthermore, I think Guerrero is largely correct about the conditions for culpable ignorance. But his discussion of abortion is either viciously circular, or if DKDK is weakened to avoid such a charge, his conclusion is in tension with his discussion of culpable ignorance.

We are getting closer to answering the question in what way might DKDK differ from the notion of moral risk canvassed in this book. Inching toward such an answer requires canvassing one last problem with Guerrero's discussion of DKDK as applied to abortion. The key problem is that on my view, it does not matter so much whether the mother has first or second order knowledge about the moral status of the unborn child. What matters is whether the OB/GYN performing the abortion can justifiably claim that she knows that she doesn't know. Yet, very few focus on the epistemic situation facing the abortion doctors who are the ones who perform an abortion. Clearly an abortion doctor is *culpably* ignorant of the moral status of the preborn baby. We know this because they themselves give voice to such moral tension of doing an abortion. Since this is a crucial point, consider at length what one abortion doctor observes.

> There is violence in abortion, especially in second trimester procedures. Certain moments make this particularly apparent, as another story from my own experience shows. As a third-year resident I spent many days in our hospital abortion clinic. The last patient I saw one day was 23 weeks pregnant. I performed an uncomplicated D&E procedure. Dutifully, I went through the task of reassembling the fetal parts in the metal tray. It is an odd ritual that abortion providers perform – required as a clinical safety measure to ensure that nothing is left behind in the uterus to cause a complication – but it also permits us in an odd way to pay respect to the fetus (feelings of awe are not uncommon when looking at miniature fingers and fingernails, heart, intestines, kidneys, adrenal glands), even as we simultaneously have complete disregard for it.
>
> Then I rushed upstairs to take overnight call on labour and delivery. The first patient that came in was prematurely delivering at 23–24 weeks. As her exact gestational age was in question, the neonatal intensive care unit (NICU) team resuscitated the premature newborn and brought it to the NICU. Later, along with the distraught parents, I watched the neonate on the ventilator. I thought to myself how bizarre it was that I could have legally dismembered this fetus-now-newborn if it were inside its mother's uterus – but that the same kind of violence against it now would be illegal, and unspeakable. Yes, I understand that the vital difference between the fetus I aborted that day in clinic, and the one in the NICU was, crucially, its location inside or outside of the woman's body, and most importantly, her hopes and wishes for that fetus/baby. But this knowledge does not change the reality that there is always violence involved in a second trimester abortion, which becomes acutely apparent at certain moments, like this one (Harris, 2008, pp. 76–77).

The point of this extended quotation of course is that abortion doctors themselves satisfy DKDK. If they claim not to know the moral status of what it is they kill, they are culpably ignorant according to Guerrero's criteria. Guerrero discusses to great effect the ancient slaveholder case according to which they ought to have known better. Likewise, those who kill developing human beings and assemble the very pieces of their dismembered bodies should know better. Location, as Harris herself observes, should be viewed with suspicion as demarcating the moral community. So, even if DKDK does not apply to a pregnant woman considering an abortion, it should—on Guerrero's view—apply to an abortion doctor. Since the abortion doctor is the one who performs the action anyway, that DKDK applies to him or her is what matters.

We can now see how Guerrero's DKDK is different from the notion of moral risk advocated for here, namely premise (B). (B) is not framed with reference to any particular agent but refers instead to the objective epistemic features pertaining to (pk). In fact, it is quite easy to motivate principles such as DKDK when indexed to specific agents, for agents can think any number of things (mistakenly or unmistakenly) which should by their own lights give them reasons for not acting. Because (B) is indexed to the objective dialectical features pertaining to (pk), a detailed analysis of its justifications is required. This latter feature is what distinguishes the notion of moral risk advocated here from other principles such as the precautionary principle, uncertaintism, and pragmatic encroachment. Knowing whether moral risk provides grounds for diffidence requires knowing the moral consequences of acting on (pk) if it were false, *in relation to its justifications*. Because (B) is not indexed to any specific agent as DKDK is, the conclusion that follows given the discussion so far is that *no one* may act on (pk)—if (C) is also true.

To premise (C) we turn to next.

8.3 The Asymmetry of Moral Risk

Chapter 2.3 closed by noting that all should agree that the moral risks are asymmetrical in the sense that the proponent of (pk) bears the greater risk—namely, liberty interests are constrained by concerns about commutative justice. This is a controversial claim indeed as contemporary philosophy has long held that the moral risks pertaining to abortion (and to a lesser extent euthanasia) are entirely on the side of those who think that such actions are impermissible. Being wrong about ~(pk) and acting on it would have drastic moral (and social) consequences.

A tempting but meretricious defense of the claim that the risks are on the side of ~(pk) is to focus attention on what one actually believes versus what might be possible. Christopher Kaposy explains that even in the setting of uncertainty about whether the fetus is a person or that abortion is wrong, one should favor the pro-choice position. The reason: "The indisputable rights and interests of women cannot be outweighed by the disputable and indeterminate rights of the fetus" (2010, p. 154). The idea here is a permutation of the position that one ought to act strictly on what one actually believes about the morality of an action. Notice that for Kaposy, there is no need to appeal to one's second-order knowledge. Considerations of what might be the case may be set aside. Considering what might be the case include not just that the action might turn out to be seriously wrong, or that my thinking on the issue may have been faulty, but even considerations to the effect that I might change my mind in the future. Richard Foley has his readers entertain a case of intra personal disagreement according to which we typically change our opinions and beliefs throughout our lifetime. Some changes are small and some involve major conversions. But the fact that we change our beliefs over time is about as consistent as that we change cells over time. Now imagine your future self looking back on your current self vis-à-vis your belief on abortion. Suppose you have converted, i.e. you now hold ~(pk). Given Kaposy's reasoning, we may set that possibility aside as well. But that claim is implausible.

This position is problematic for several reasons: it is a position that precludes any consideration of moral risk, precaution, pragmatic encroachment, uncertaintism, and possibly other positions within the family of decision theory in the setting of uncertainty.

What is more problematic yet is that in picking sides he is assuming (pk). He mentions the "indisputable rights and interests of women." A right to what? An interest in what? Clearly, Kaposy means a right to and interest in procuring an abortion: "it is in the interests of women to be able to terminate unwanted pregnancies" (2010, pp. 153–154). What he thinks is "indisputable" then, just is (pk).[6] It is easy to see how any permutation of Kaposy's claim here is also circular. The essential ingredient is that one would be preferring the putatively *known* (pk) to the unknown ~(pk).

Not only is this claim circular, but in the context of Kaposy's argument, it is contradictory. This claim contradicts the very premise of his paper which is that "any one position on the moral standing of fetuses cannot be established decisively" (Kaposy, p. 139). And just a few lines up from the quotation on indisputable rights, he states that "It is not meaningful to lay odds on the question of fetal moral standing because laying odds presupposes that there is a determinate answer to the question…" (2010, p. 153). If the moral status of the fetus is indeterminate, that is, if we cannot be certain about his or her moral status, how can it be "indisputable" that the fetus may be killed? One cannot be both certain and indeterminate about X.

A different argument to the same conclusion is offered by Roger Wertheimer. He argues in (1971) that the default position should be (pk). To quote in full, he says,

> The existence and powers of the state are legitimated through their rational acceptability to the citizenry, and it would be irrational for the citizens to grant the state any coercive power whose exercise could not be rationally justified to them. Thus, the state has the burden of

---

[6]The term indisputable in its ordinary usage typically refers to what is agreed upon by all parties to a dispute. Clearly, Kaposy cannot mean that in this context since it is quite disputable whether abortion is permissible. I interpret him to be making an epistemic claim to the effect that the reasons and justification for (pk) make (pk) certain or indisputable.

> proving that its actions are legitimate. Now, without question, the present abortion laws
> seriously restrict the freedom and diminish the welfare of the citizenry (1971, p. 94).

Though framed as what epistemic obligations the state has to its citizens, one might translate this initial argument to moral deliberation on abortion or euthanasia roughly as follows. A desperate mother requests an abortion from an OB/GYN. An OB/GYN's refusal to perform the abortion should be "rationally acceptable" by the mother. If it were not, the OB/GYN would exercise coercive power that is not rationally justified to the one who is coerced, i.e. the mother. Therefore, the OB/GYN bears the burden of proof in so far as her refusal restricts the mother's freedom. Wertheimer concludes saying that,

> the state has the burden of showing that such a law [a restrictive abortion law] is necessary
> to attain the legitimate ends of the state. But the social costs of the present abortion laws [he
> is writing prior to *Roe v. Wade*] are so drastic that only the preservation of human lives could
> justify them. So to justify those laws *the state must demonstrate that the fetus is a human
> being* (1971, p. 94, emphasis added).

This initial statement of the argument is rather underdeveloped as it stands. First, if it is not a human being, then what is it? Does the human embryo start out as a caterpillar and then turn into a human being at birth? Framing the argument in the way Wertheimer states it must set aside a large swath of embryological evidence or be stuck with the absurd claim that the developing human being is not a human being.

Second, Wertheimer is inattentive to the social costs of allowing the killing and dismemberment of human beings. The social costs of allowing a society to kill its most vulnerable members strikes me as enormous if it were true that abortion is a killing of vulnerable human beings. Wertheimer's risk argument must not only presuppose (pk) but view any argument against it or for ~(pk) as having no epistemic weight—a presupposition inconsistent with his own analysis of the dialectic.

Wertheimer is not clear on what exactly the drastic social consequences are, but as with most social policies, such consequences are hard to determine prior to their implementation. In any case, here's one consequence: if abortion laws were restrictive, members of society would have to tailor their sexual choices to avoid unplanned pregnancies. Is this consequence "drastic"? Are we to take seriously the idea that this cost is more drastic than the cost of unjust killing?

In any case, Wertheimer's argument is that restrictive abortion laws constrain one's liberty and such a constraint bears the burden of proof. Wertheimer suggests that the arguments he has canvassed do not demonstrate or prove that the fetus is a human being. On this point, the reverse is also true. Throughout his article blips of skepticism shine through with such statements as "[i]n the abortion case my instincts are similar but *shakier*" (Wertheimer, 1971, p. 93). And "does accepting the liberals reply scotch all further argument? I think not" (Wertheimer, 1971, p. 93). Both quotations immediately precede his risk argument outlined here.

He has also failed to develop his argument in the following sense. It is unclear what his epistemic standards are for providing a demonstration for any claim. In a way, the higher the standard he sets for the pro-life argument, the stronger his own argument must be for why liberty interests alone set the burden of proof. He certainly does not *demonstrate* this. Consistency on such matters should be maintained across arguments.

A more careful presentation of an argument that the moral risks are entirely on the side of those who believe (and act on) ~ (pk) comes from Thomson (1995). Her argument is succinctly stated as follows.

> First, restrictive regulation [of abortion] severely constrains women's liberty. Second, severe constraints on liberty may not be imposed in the name of considerations that the constrained are not unreasonable in rejecting. And third, the many women who reject the claim that the fetus has a right to life from the moment of conception are not unreasonable in doing so (Thomson, 1995).

The value of liberty paired with some notion of 'not unreasonable' are taken by Thomson to set the dialogical burden on the pro-life argument. If Thomson is right, any pro-life argument for which it is reasonable to reject one of its premises is enough to permit someone to act as if the argument is incorrect. I hope to show that her argument is subject to reasonable rejection.

Because I refer to specific premises in what follows, enumerating her argument is necessary.

I. Restrictive regulation of abortion severely constrains women's liberty.
II. Severe constraints on liberty may not be imposed on the basis of arguments that *the constrained* might reasonably reject.[7]
III. Some of those who might be constrained reasonably reject arguments for the restriction of abortion.
IV. Therefore, restrictive regulation of abortion should not be imposed.

The argument has a meretricious attraction to it. It sounds like it is a good argument. But there are multiple and, I argue, individually sufficient reasons for rejecting it.

First, there is an equivocation on the notion of liberty between premises (I) and (II). The value of liberty referred to in (I) is the putative value of making *this choice* for abortion—restrictive regulation would prohibit making this choice. Conversely, premise (II) is plausible only when "liberty" is taken to refer to one's *capacity* to

---

[7] The change from 'not unreasonable in rejecting' to 'reasonable to reject,' or its cognates, is motivated by charity. Not being unreasonable is a weak epistemic state. Coherence is not even a requirement for 'not being unreasonable' in believing B since the latter only requires that B is not incompatible with my network of other beliefs N. But my belief B may not be supported by them either—as would be involved if B were coherent with N. Not being unreasonable in believing B, then, is compatible with believing B for no reason whatsoever—just so long as one has no reasons against B. On this understanding, the dimwitted Nazi can be 'not unreasonable' in believing the stupid things she does. Such an epistemic state would render premise (II) initially implausible. The premise would permit rejecting seatbelt laws but for no coherent reasons that support such a rejection. In the interest of charity, then, Thomson must have in mind a stronger notion such as being reasonable in rejecting B which requires some support relation with a coherent network of other beliefs N. Thus, even if there are reasons for rejecting restrictive abortion policies, I still wish to argue that permissive abortion policies should not be presumed. As such, I am setting the bar higher for my argument not Thomson's.

make choices. I walk into a restaurant pining for their special. The waiter informs me that they are out of their special. Though a choice is taken away, my capacity for making choices is retained and remains a valuable capacity that I possess qua person.

Now consider the counterintuitive ring of understanding premise (II) as:

- severe constraints on one's specific choices may not be imposed on the basis of arguments the constrained might reasonably reject.

We do this all the time and justifiably so. Even those who are constrained may reasonably accept such constraints (Callahan, 2004). Examples abound such as the regulations governing human subjects research, especially on infants and children (see 45 CFR 46 Sub-part D). Environmental regulation constrains our choices, but those who are constrained (i.e. the energy industry and those who own Hummers) might have coherent reasons for rejecting these constraints. It certainly doesn't follow that they are off the hook simply because they don't see such constraints as justified.

Rephrasing premise (I) is even less intuitively persuasive. Consider,

- restrictive regulation of abortion severely constrains women's *capacity* to make choices.

Whatever restrictive regulation of abortion does, it certainly does not impugn one's *ability* to make decisions. So, when Thomson claims that restrictive abortion policies "severely constrain women's liberty," a categorical error is being made between (I) and (II). Two different values are being conflated: the value of having free will at all, namely, the capacity to make choices and the value of choosing this specific option.

But more needs to be said about how the two values, the value of a specific choice and the value of the capacity to make choices, can come apart. Consider a pregnant woman and her liberties and freedoms. Presumably, simply being pregnant does not "severely constrain one's liberty," at least not in a way that would ground some moral prohibition on getting pregnant. Granted, parents have to make different choices than non-parents. But parents clearly retain the capacity to make such decisions. And, although the specific choices are different between parents and non-parents, they are no less valuable. So, if premise (I) is plausible, it is likely that premise (II) is not, because premise (II) is plausible only when we consider the value of liberty per se, qua capacity, and not the value of this or that specific choice, and vice versa if (II) is plausible.

Reflection on the examples of environmental or human research regulations invites the following problems for Thomson's argument. Premise (II) refers to the women who might be constrained by restrictive abortion policies. And it says *of them* that they might be reasonable in rejecting pro-life arguments. It is not clear what counts as being reasonable for Thomson: Is it coherence, justification with no known defeaters, reliably formed belief etc.? Aside from these details, one can *understand* why a woman faced with a crisis pregnancy would find abortion as an attractive option. If reasonability is defined as means-end reasoning, those facing crisis pregnancies would be reasonable in considering abortion. But this is not what

Thomson needs for her argument. I might fail to appreciate the weight of the reasons for a coercive taxation policy—because I have an interest in keeping my own hard-earned dollars. But that would hardly justify the state in not imposing such a taxation policy. Premise (II) needs to say that pro-life arguments can be rejected based on reasons that are objectively plausible. To motivate why, imagine the following parody of Thomson's argument understood without this objective feature.

I′. Restrictive regulation of slavery severely constrains slaveowners' liberty.
II′. Severe constraints on liberty may not be imposed based on arguments that the constrained might reasonably reject.
III′. There are those who reasonably reject arguments for the restriction of slavery.
IV′. Therefore, restrictive regulation of slavery should not be imposed.

All I have done in constructing this parity is swap out 'abortion' for 'slavery'. (Notice also just how unpersuasive the argument would be if we included the original 'not unreasonable in rejecting' criterion.)

Suppose Thomson understands 'reasonable rejection' objectively in the sense that opponents to restrictive abortion policies have objectively good reasons for their position. So, her argument from risk must include the following conditional or one trivially different.

- (O1) If there are objective good reasons for *opposing* restrictive abortion measures, abortion should be allowed. ("Allowed" because in premise (II), reasonable rejection is said to justify non-imposition.)

The chief problem with this conditional is that it does not consider whether there are objective reasons for *endorsing* restrictive abortion policies. The problem is that anyone can think up reasons for a position p, but if there are potent counter-vailing reasons the consequent in (O1) does not follow. Suppose that, if true, the idea that human beings maintain their ontological identity through development qua persons is a good reason for restricting abortion access. Conversely, if true, the idea that unborn human beings are not persons would be a reason for opposing restrictive abortion measures. Both strike me as good reasons. What is not at all plausible, however, is to focus on only one or the other reason as sufficient for permitting (or preempting) the act in question. It is in *this* setting of good objective reasons on *both sides* that one mounts a moral risk argument. So, we should understand the key conditional as follows.

- (O2) If there are good objective reasons for both opposing and endorsing restrictive abortion policies, abortion should be allowed.

Once we have captured the dialogical space on abortion accurately, (O2) is hardly apparent. If the objective reasons are good ones on both sides, it simply does not follow that abortion should still be allowed. (Consider the counterintuitive ring of a hunting version of O2, or a slavery version of (O2): if there are good objective reasons for opposing and endorsing restrictive hunting practices/slavery, unrestricted hunting practices/slavery should continue).

That pro-life arguments inherit the initial dialogical burden would follow only if the liberty involved in exercising *this* choice (the choice to have an abortion) is a value that outweighs all other value considerations (i.e. the values referred to by the arguments for restrictive policies). But that axiological claim is neither obvious nor true. All parties agree that killing a human person is worse than violating someone's liberty interest—no one justifies killing children simply because the mother and father must alter their behavior and choose to care for the born child. Even if they really want to do other things with their time, caring for their born children is still obligatory. Failures to care, no matter how strong the liberty interest might be, are viewed as child abuse.

Even if this last observation is contestable, how would one argue for the claim that liberty interests hold a privileged position? It seems we have to appeal to moral intuitions, but when we do we are back to the examples just alluded to: We do not endorse the killing of our children even if taking care of them severely restricts our individual liberties. The value of our liberty is circumscribed.

Importantly, the value of that liberty does not itself justify a moral difference between born and unborn children. Having a liberty interest does not justify the ontological/moral status of the child whether born or pre-born. What this means is that the value of liberty is a value that is logically posterior to having settled the moral status of the pre-born child. So, the life of one's child functions as a value the elimination of which can be considered a significant error; and that value determines in part the value of one's choices. To be clear, violations of one's liberty can be considered a moral cost to getting an issue wrong. My argument here is that it cannot be a symmetrical cost to the moral cost of unjustly killing an innocent child.

As is anticipated in Chap. 2, I conclude here that disputants to the issues of abortion and euthanasia *agree* on what values are important and which position would, if false, involve a greater harm in acting on it. For abortion, all disputants agree that a woman's liberty interests are important values. All disputants agree that unjustly killing persons is a significant harm that ought to be avoided. And finally, all disputants agree that our liberty interests bend or are subordinated to the value of persons. If parents no longer want their born children, we don't accept that they can neglect, abuse, or eliminate their children. Their wants and interests are valuable, but not as valuable as protecting the good of the child. As such, everyone can agree that *if* opponents to abortion are correct, there is no morally relevant difference between born and pre-born children. And so, proponents of abortion can agree that *if their position is false* (if (pk) is false), then acting on it would involve a grave injustice; one more serious than if ~(pk) is false.[8]

For euthanasia, the asymmetry can be stated much more succinctly. As the discussion of Sumner and several others made clear, the moral cost of not performing euthanasia in many cases is that the patient is left to suffer longer. As Chap. 7 made

---

[8] The important reminder here is that the project of distributing which position bears the moral risk is different from the project of explaining how to manage it. As such this paragraph is not implicitly advocating for some notion of uncertaintism as understood by Boonin.

clear, however, there are palliative measures that are quite effective. Even so, palliative, deep sedation is effective in suppressing someone's suffering (Sulmasy, 2018; Patel et al. 2019; and Bhyan et al., 2024) and it does so without the moral risk of an intentional killing.

8.4   The Argument in Detail

It may be granted that in the main, the argument at the beginning of the chapter (A)-(D) is valid, but not yet sound. That is to say, one might grant that there are de jure defects with the argument for (pk) but continue to think that it enjoys the better argument than, for example, the modal argument for ~(pk). So, let's descend to details to see if this claim is plausible.

Consider a recurring motif in Chap. 4 to the effect that the hylomorphic account of the person can accommodate the intuitions of the transplant cases, but it understands the person to be a substance, an entity with a rational nature. Consciousness, on the hylomorphic view, is a capacity of the person, but is not itself the person. This is contrasted with the functional view according to which being a person entails having exercisable psychological capacities. The differences between these views is something like this:

> (Actualism) = You and I are persons at t *iff* you and I *can* exercise psychological capacities at t.
> (Non-Actualism) = S is a person at t *iff* S is an individual with a rational nature at t.[9]

Suppose, as was argued in Chap. 4, that the typical thought experiments used to motivate the functional view of the person do not force one to decide between actualism and non-actualism. The reason, again, is that the latter can accommodate the intuitions on the transplant cases, dicephalic twin case, and the body-transplant case.

Suppose that (Actualism) is required to justify (pk) on the no-person strategy. Actualism is under-motivated, meaning that alternative theories can accommodate the relevant metaphysical intuitions putatively favoring (Actualism). Both theories are equiprobable given the intuitive data. It follows that (Actualism) may (or may not) be true. There is no reason *for* it in the sense that it enjoys exclusive justification. Now, crucially, replace (pk) with (Actualism) since the justification for a claim that bears moral risk also bears that risk—especially if the justification is required. The moral risk in being wrong about (Actualism) is that if it were false and acted upon, one would kill a person. Since, as 8.3 argued, everyone agrees that this would be a serious moral cost that is asymmetrical to the cost of violating one's liberty

---

[9] In Chap. 4, I contrasted what I am here calling Actualism with the hylomorphic account. But it can be any view of the person that does not reify the psychological capacities into a separate being, such as animalism. So, non-Actualism should not be understood simply as the negation of Actualism (but it would logically entail the negation of Actualism).

interests, everyone should agree that one should not act as if (Actualism) were true. To put the previous train of thought into syllogistic form:

(1) Actualism → (pk).
(2) There is no reason favoring Actualism over non-Actualism. (This is the problem of underdetermination).
(3) If Actualism → (pk) and (pk) bears a greater risk in being wrong than ~(pk), then Actualism bears a greater risk in being wrong than non-Actualism. (Instantiation of Premise B).
(4) (pk) bears a greater risk in being wrong than ~(pk). (Premise C).

From (1) and (4) we can infer the consequent in (3):

(5) Therefore, Actualism bears a greater risk in being wrong than non-Actualism.

(2) and (5) satisfy the antecedent conditions of (B). So,

(6) If (B)—specified here to Actualism—then Actualism should not be acted upon.
(7) (B)—specified to Actualism.
(8) Therefore, Actualism should not be acted upon.

One can recapitulate this procedure for each hinge proposition for the respective arguments and purported de jure defects.

So, return for a moment to the argument for (pk) specified to euthanasia. Chap. 6 revealed that the argument for (pk) required arguing that death could benefit a person. The continuum argument concludes that as one travels 'down' in functionality with death as a terminus, no one would say that the patient benefits. Intuitively, nothing changes when one arrives at the terminus. But suppose that your intuitions are shaky on this point. We may isolate out two competing claims as above:

> (Existentialism) = As S loses health functioning in progressive stages, S is not benefited by each loss; and this includes loss of S's existence.
>
>   (Non-existentialism) = As S loses health functioning in progressive stages, S is not benefited by each loss *except* the loss of existence.

Non-existentialism is required, on this hypothesis, to justify (pk). But Non-existentialism is obviously false; or if you prefer, counter-intuitive. At the very least, there is no reason *for* it. As with (Actualism), replace (pk) with (Non-existentialism) since the latter now bears the same risk in being wrong about (pk)—because it is, on this hypothesis, required justification. The moral risk is that one would unjustly kill an innocent and vulnerable human person. As above, we can construct the following argument:

(9) Non-existentialism → (pk).
(10) There is no reason for favoring non-existentialism over existentialism. (To transplant one of the points made in Chap. 6, those who think that there is an exception for death itself already think that such patients may be killed. A circular justification for the former does not provide a reason for favoring it to the latter.)

(11) If Non-existentialism → (pk) and (pk) bears a greater risk in being wrong than
      ~(pk), then Non-existentialism bears a greater risk in being wrong than exis-
      tentialism. (Instantiation of Premise B). (Again, as pointed out in Chap. 6, if
      the problem is the suppression of suffering, this can be accomplished without
      intentional killing, viz., terminal sedation. So, very little moral cost attaches to
      existentialism).[10]

(12) (pk) bears a greater risk in being wrong than ~(pk). (Premise C).

From (9) and (12) we can infer the consequent in (11):

(13) Therefore, non-existentialism bears a greater risk in being wrong than
      existentialism.

(10) and (13) satisfy the antecedent conditions of (B). So,

(14) If (B)—specified here to non-existentialism—then existentialism should not
      be acted upon.

(15) (B)—specified to non-existentialism.

(16) Therefore, non-existentialism should not be acted upon.

Of course, readers will place emphases on different syllables of an argument. So,
one might think that a crucial feature that makes loss of existence a benefit is that it
ends the patient's suffering. Chapter 6 highlighted three points in this regard. First,
there are other cases in which a vulnerable person is suffering, but that is not a feature that would justify an exception to the prohibition on intentionally killing innocent persons. Our intuitions are at best mixed. As above, there is no reason *for* the
claim that suffering is a feature of a person that justifies killing him or her. Second,
there are other ways to end suffering, namely terminal sedation. If terminal sedation
is done without the intention to dehydrate the patient to death, it bears little moral
risk. Third, all agree that suffer*ing* should be managed. Because of the obvious categorical difference, it does not follow that one is justified to kill the suffer*er*.

As above, we can isolate out the claim that suffering is a feature of the sufferer
that justifies killing the person; and its corollary, suffering is not a feature that justifies killing an innocent person. Given that our institutions are mixed, there is no
reason for the claim that suffering is a feature that justifies killing. Given the risk in
being wrong about (pk) in the setting of low risk alternatives (i.e., terminal sedation,
or dignity therapy), one ought not to act on (pk).

The pattern is easy to see, and it can be iterated for each justification and that
justification's de jure defect. Does the same conclusion follow for the modal argument (Chap. 5)? The de jure defects were twofold: the argument assumes that it is
wrong to kill persons; and the argument assumes that the abortion survivor should
say that "*I* was almost killed." I conceded that the argument falls prey to the first, but

---

[10] One might say that the moral cost is the violation of one's autonomy. But this too was addressed
in Chap. 6. The contrasting positions would be the view of autonomy's worth as anchored to intra-
life decisions or uncircumscribed to include contra-life decisions. Again, the same argumentative
pattern as noted here can be recapitulated for such 'joints' in the dialectic.

I also argued that the putative de jure defect in question suffers from its own de jure defect. While it is true that I do not possess the same set of rights as I did when I was an infant, the one right I do share with my infant is the right not to be intentionally killed. Boonin has not argued that the right not to be killed is one of those rights that covaries with development.[11]

To the second, the defect suggests that I have assumed precisely what is at stake. Maybe the modal argument is not *assuming* that the abortion was done on Sarah, but is *revealing* it. In any case, let's isolate out the respective claims at this point in the dialectic. Consider an abortion survivor describing the abortion performed on her. From the first person perspective she can say:

> (Identity) = I was almost killed by an abortion. Or,
>   (Non-identity) = I almost did not come into existence because of the abortion.

One reason *for* (Identity) is that we have no other way to account for *her* injuries—many abortion survivors are permanently injured by the attempted abortions performed on them. And it is patently obvious that *they* were injured by the abortion attempt. Second, my interlocutors share the same intuitive pull of (Identity) at least when considering prenatal harm cases. The only reason for (Non-identity) is if one assumes a functional view of persons. But there is a reason why Chap. 5 occurs after Chap. 4. The result of Chap. 4 undercuts assuming that a functional view of persons is a superior way of understanding persons. With this in mind, replace, as above, (pk) with (Non-identity) since the latter is required to preserve (pk) in the face of an argument against it. Since the moral risks are asymmetrical, (pk) should not be acted upon. Following the pattern of the above arguments (1–8) and (9–16) it would follow that Non-identity should also not be acted upon.

What is more, the second objection is quite fundamentally, underdetermined. Assume that (Non-identity) depends on the truth of the functional view, or a view trivially different. The procedure for motivating the functional view, however, is to rely on our metaphysical intuitions on cases. That methodological assumption is the same one used to consider cases of Transworld identity between an abortion survivor, and the nearest possible world in which she does not survive. The intuition is still that from the abortion survivor's first-person perspective, (Identity) is true. Because the modal argument executes a methodologically similar procedure to motivate its conclusions it can function itself as a de jure defect for the no person argument for (pk). The defect in question is a justification defect satisfying condition (ii) according to which there are similar justifications but for opposite conclusions. Importantly, Chap. 5 pointed out a relevant asymmetry here namely that the intuitions on the Transworld identity cases gives direct intuitive evidence for ~(pk),

---

[11] In fairness to Boonin he does argue that consciousness is a value. What he does not argue for, however, is that it is the only value. What he needs to argue is that being an entity that will of biological necessity develop a capacity for consciousness is not also a valuable *being*. Interestingly, he appears to admit that that entity is me. On this point, Boonin rightly resists identifying persons with their capacities. His only error is locating our worth only in the *developed* capacities we have.

whereas the brain transplant thought experiments require supplementary explanations.

The conclusion reached in this book is that there is no justification for (pk) that does not suffer from comparatively worse de jure defects than those affecting arguments for ~(pk). Second, the moral risk in being wrong is more serious for (pk) than it is for ~(pk). Given the inference in premise (B) it follows that one should not act on (pk). Notice that the conclusion is not that (pk) has no justification whatsoever. It is not that (pk) is false. It is not, even, that no one should believe (pk). It is that the present development of philosophical arguments for (pk) are such that they suffer from comparatively worse defects as those affecting ~(pk) and the moral risks are asymmetrical. My interlocutors can continue believing (pk), I have offered no argument against that here. They can even continue to think that (pk) is true. My argument is that (pk) should not be acted upon.

8.5 Conclusion

One may still have the lingering thought that an agent may act on her own conscience even if that conscience may be wrong. It looks like I am recommending precisely what I said I found plausible about reflective equilibrium—such a method of moral inquiry is the only rational method because it does not ask the agent to believe something she thinks is false or implausible. Does my conclusion just stated above threaten rationality? Would it be irrational for an abortion doctor to say to a woman in a crisis pregnancy that "I refuse to do this abortion because I know of some pro-lifers who disagree with me, and there is a risk in wrongdoing if I do the abortion"? It seems so.

But that is not precisely what I am recommending in this book. It would not be silly or irrational for the OB/GYN to say that though he believes (pk), "after considering the best arguments for the permissibility of abortion and plumbing their respective de jure defects, I have enough reason to be diffident that it is permissible; diffident enough to not be justified in acting on my belief given the risk in being wrong about it." There doesn't seem to be anything silly about such a response once it fully represents my position. It is true that we make our judgments and we act on them as we see fit. But it is equally sensible to assume that we conduct our epistemic lives with epistemic humility and intellectual care. We have to be responsive to reasons.

Let's consider some parallel cases where it is quite obvious we would want the agent to be more diffident. Suppose the setting is on a plantation during the *Post-Garrison* era of the South and suppose that the plantation owner's family is begging him (call him Jones) to get another slave. Is it so silly for Jones to respond "well, I know of these abolitionists who seem good-hearted and they make some good points against the institution of slavery. And there is a tremendous moral risk in getting my own judgment wrong. So I'm not going to get a slave." Granted, we would like the slave owner to come right out and reject slavery. The point is to highlight that for the generic reasons canvassed in this book, his diffidence is not irrational.

Consider another setting in which a soldier is asked to torture a POW. Suppose I'm the soldier and I respond "I just read some arguments from apparently good-hearted people familiar with the complexities of war and espionage who say that torturing POWs would still be wrong. I do not believe it, but I don't appear to have an epistemic edge on their reflections. The reasons for its permissibility suffer de jure defects of some sort or other, and there is a risk in being wrong. So I refuse." There is nothing irrational about this response. As with the previous case, we may want the soldier to come right out and reject torture, but that is not the point. The point is whether the generic reasons offered in this book can justify diffidence in parallel cases. I suggest that they clearly can.

Bibliography

Alexander, L. (1949). Medical science under a dictatorship. *New England Journal of Medicine, 241*(2), 39–47.

Almeida, M.J. (2004). Marginal Cases and the Moral Status of Embryos. In Stem Cell Research. Biomedical Ethics Reviews. Eds. Humber, J.M., Almeder, R.F. (pp. 23–39). Humana Press.

Alston, W. P. (1985). Concepts of epistemic justification. *The Monist, 68*(1), 57–89.

Alston, W. P. (1991). *Perceiving God: The epistemology of religious experience*. Cornell University Press.

Alston, W. P. (1993). *Perceiving God: The epistemology of religious experience*. Cornell University Press.

Alston, W. P. (2005). *Beyond "justification": Dimensions of epistemic evaluation*. Cornell University Press.

AMA (American Medical Association) Code of Ethics. https://code-medical-ethics.ama-assn.org/. Accessed 31 July 2024.

Anscombe, E. (2000). *Intention*. Harvard University Press.

Anscombe, G. E. M. (1958). Modern moral philosophy. *Philosophy, 33*(124), 1–19.

Armstrong, D. M. (1980). *A theory of universals: Volume 2: Universals and scientific realism*. Cambridge University Press.

Arras, J. (2006). The way we reason now: Reflective equilibrium in bioethics. In *The Oxford handbook in Bioethics*. Ed. B. Steinbock (46–71). Oxford University Press.

Arras, J. (2010). Physician-assisted suicide: A tragic view. In L. Vaughn (Ed.), *Bioethics: Principles, issues, and cases* (pp. 565–579). Oxford University Press.

Aru, J., & Bachmann, T. (2015). Still wanted—The mechanisms of consciousness! *Frontiers in Psychology, 6*, 5.

Audi, R. (2001). *The architecture of reason: The structure and substance of rationality*. Oxford University Press.

Audi, R. (2013). *Moral perception*. Princeton University Press.

Austin, C. J., & Marmodoro, A. (2017). Structural powers and the homeodynamic unity of organisms. In W. M. Simpson, R. C. Koons, & N. J. Teh (Eds.), *Neo-Aristotelian Perspectives on Contemporary Science* (pp. 183–198). Routledge.

Austriaco, N. P. G. (2004). Immediate hominization from the systems perspective. *The National Catholic Bioethics Quarterly, 4*(4), 719–738.

Austriaco, N. P. G. ed. (2024). Creation through evolution: New perspectives from Thomistic philosophy and theology. CUA Press.

Austriaco, N. P. G., Brent, J., Davenport, T., & Ku, J. B. (2019). *Thomistic evolution* (2nd ed.).

Bach, J. R. (2003). Threats to "informed" advance directives for the severely physically challenged? *Archives of Physical Medicine Rehabilitation, 84*(Supplement 2), S23–S28.

Baker, L. R. (1995). Need a Christian be a mind/body dualist? *Faith and Philosophy, 12*(4), 489–504.

Baker, L. R. (2000). *Persons and bodies: A constitution view*. Cambridge University Press.

Baker, L. R. (2005). When does a person begin? *Social Philosophy and Policy, 22*(2), 25–48.

Bauernfeind, G., Scherer, R., Pfurtscheller, G. et al. (2011). Single-trial classification of antagonistic oxyhemoglobin responses during mental arithmetic. *Medical Biological Engineering and Computing 49*, 979–984.

Beauchamp, T. L., & Childress, J. F. (2001). *Principles of biomedical ethics* (5th ed.). Oxford University Press.

Beckwith, F. J. (2007). *Defending life: A moral and legal case against abortion choice*. Cambridge University Press.

Beckwith, F. J. (2011). The human being, a person of substance: A response to Dean Stretton. In Persons, Moral Worth, and Embryos. Ed. Stephen Napier. (pp. 67–83). Springer Publishers.

Bennett, M. R., & Hacker, P. M. S. (2003). *Philosophical foundations of neuroscience*. John Wiley & Sons.

Bennett, M. R., & Hacker, P. M. S. (2008). *History of cognitive neuroscience*. John Wiley & Sons.

Bergelson, V. (2008). Autonomy, dignity, and consent to harm. *Rutgers Law Review, 60*, 723–736.

Bergmann, M. (2004). Epistemic circularity: Malignant and benign. *Philosophy and Phenomenological Research, 69*(3), 709–727.

Bertalanffy, L. V. (1949). *Problems of life*. John Wiley and Sons.

Bhyan, P., Pesce, M. B., Shrestha, U., & Goyal, A. (2024). Palliative sedation in patients with terminal illness. In *StatPearls [Internet]*. StatPearls Publishing.

Bigelow, J., & Pargetter, R. (1988). Morality, potential persons and abortion. *American Philosophical Quarterly, 25*(2), 173–181.

Bishop, M. A. (1999). Why thought experiments are not arguments. *Philosophy of Science, 66*(4), 534–541.

Blackshaw, B. P. (2019). The impairment argument for the immorality of abortion: A reply. *Bioethics, 33*(6), 723–724.

Blanshard, B. (1944). Current strictures on reason. *Proceedings and Addresses of the American Philosophical Association, 18*, 345–368.

Blanshard, B. (1961). *Reason and goodness*. London: George Allen and Unwin Publishers.

Blum. Lawrence. (1994). *Moral perception and particularity*. Cambridge University Press.

Bonjour, L. (2009). Internalism and externalism. In P. K. Moser (Ed.), *The Oxford handbook of epistemology* (Oxford handbooks. (online edn, Oxford Academic, 2 September 2009)). https://doi.org/10.1093/oxfordhb/9780195301700.003.0008

Boonin, D. (2003). *A defense of abortion*. Cambridge University Press.

Boonin, D. (2019). *Beyond Roe: Why abortion should be legal--even if the fetus is a person*. Oxford University Press.

Braddock, M. (2024, August). Do not risk homicide: Abortion after 10 weeks gestation. *The Journal of Medicine and Philosophy, 49*(4), 414–432.

Brahms, J. (2019). Blood donation and bodily rights arguments. https://www.youtube.com/watch?v=YmBrUcpOxDw. Accessed 21 June 2024.

Brake, E. (2010). Willing parents: A voluntarist account of parental role obligation. In D. Archard & D. Benatar (Eds.), *Procreation and parenthood: The ethics of bearing and rearing children* (pp. 151–177). Oxford University Press.

Brakman, S. V. (1994). Adult daughter caregivers. *The Hastings Center Report, 24*(5), 26–28.

Brakman, S. V. (2014). Who is a parent? *The Family in America, 28*, 349–368.

Brandt, R. (2006). The rationality of suicide. In T. Mappes & D. Degrazia (Eds.), *Biomedical ethics* (6th ed., pp. 388–394). McGraw-Hill.

Brandt, R. B. (1975). A moral principle about killing. In M. Kohl (Ed.), *Beneficent euthanasia*. Prometheus Books.

Bratman, M. (1981). Intention and means-end reasoning. *The Philosophical Review, 90*(2), 252–265.

Bratman, M. (2000). Reflection, planning, and temporally extended agency. *The Philosophical Review, 109*(1), 35–61.

Breitbart, W., Pessin, H., Rosenfeld, B., Applebaum, A. J., Lichtenthal, W. G., Li, Y., et al. (2018). Individual meaning-centered psychotherapy for the treatment of psychological and existential distress: A randomized controlled trial in patients with advanced cancer. *Cancer, 124*(15), 3231–3239.

Breitbart, W., et al. (2000). Depression, hopelessness, and desire for hastened death in terminally ill patients with cancer. *JAMA, 284*, 2907–2911.

Brewer, T. (2009). *The retrieval of ethics*. Oxford University Press.

Broackes, J. (2006). Substance. *Proceedings of the Aristotelian Society, 106*(1), 133–168.

Brock, D. W. (1992). Voluntary active euthanasia. *Hastings Center Report, 22*(2), 10–22.

Brown, M. T. (2000). The morality of abortion and the deprivation of futures. *Journal of Medical Ethics, 26*, 103–107.

Brown, M. T. (2007). The potential of the human embryo. *Journal of Medicine and Philosophy., 32*(6), 585–618.

Brown, R. (2019). Philosophy can make the previously unthinkable thinkable. *Aeon*. Available at: https://aeon.co/ideas/philosophy-can-make-the-previously-unthinkable-thinkable (accessed, 12/30/25).

Bruno, M. A., Bernheim, J. L., Ledoux, D., et al. (2011). A survey on self-assessed well-being in a cohort of chronic locked-in syndrome patients: happy majority, miserable minority. *BMJ Open, 1*, e000039. https://doi.org/10.1136/bmjopen-2010-000039

Bruno, M. A., Majerus, S., Boly, M., Vanhaudenhuyse, A., Schnakers, C., Gosseries, O., Boveroux, P., Kirsch, M., Demertzi, A., Bernard, C., & Hustinx, R. (2012). Functional neuroanatomy underlying the clinical subcategorization of minimally conscious state patients. *Journal of Neurology, 259*(6), 1087–1098.

Buchanan, A. (1996). Intending death: the structure of the problem and proposed solutions. *Intending death: The ethics of assisted suicide and euthanasia*. Ed. Tom Beauchamp (pp. 23–41). Upper Saddle River NJ: Prentice Hall Publishers.

Burke, M. B. (1994). Preserving the principle of one object to a place: A novel account of the relations among objects, sorts, sortals, and persistence conditions. *Philosophy and Phenomenological Research, 54*(3), 591–624.

Burke, M. B. (1996). Sortal essentialism and the potentiality principle. *Review of Metaphysics, 49*(3), 491–514.

Callahan, D. (1984). Autonomy: A moral good, not a moral obsession. *Hastings Center Report, 14*(5), 40–42.

Callahan, S. (2004). Pro-life feminism. In H. Gensler, E. Spurgin, & J. Swindal (Eds.), *Ethics: Contemporary readings* (pp. 275–283). Routledge.

Campbell, T., & McMahan, J. (2010). Animalism and the varieties of conjoined twinning. *Theoretical Medicine and Bioethics, 31*(4), 285–301.

Cartwright, R. L. (1960). Negative existentials. *The Journal of Philosophy, 57*(20/21), 629–639.

Cavanaugh, T. A. (2006). *Double-effect reasoning: Doing good and avoiding evil*. Oxford University Press.

Chappell, S.-G. (2004). Persons as goods: Response to Patrick Lee. *Christian Bioethics, 10*(1), 69–78.

Chappell, S.-G. (2008). Moral perception. *Philosophy, 83*(4), 421–437.

Chappell, S.-G. (2011). On the very idea of criteria for personhood. *The Southern Journal of Philosophy, 49*(1), 1–27.

Chochinov, H. M. (2004). Dignity and the eye of the beholder. *Journal of Clinical Oncology, 22*(7), 1336–1340.

Chochinov, H. M. (2023). Intensive caring: Reminding patients they matter. *Journal of Clinical Oncology, 41*(16), 2884–2887.

Chochinov, H. M., Hack, T., Hassard, T., Kristjanson, L. J., McClement, S., & Harlos, M. (2002). Dignity in the terminally ill: A cross-sectional, cohort study. *Lancet, 360*(9350), 2026–2030. https://doi.org/10.1016/S0140-6736(02)12022-8

Chochinov, H. M., Hack, T., Hassard, T., Kristjanson, L. J., McClement, S., & Harlos, M. (2005). Dignity therapy: A novel psychotherapeutic intervention for patients near the end of life. *Journal of clinical oncology, 23*(24), 5520–5525.

Christensen, D. (2007). Epistemology of disagreement: The good news. *Philosophical Review, 116*(2), 187–217.

Christensen, D. (2011). Disagreement, question-begging and epistemic self- criticism. *Philosophers Imprint, 11*(6), 1–22.

Chudnoff, E. (2013). *Intuition.* Oxford University Press.

Chwang, E. (2009). Futility clarified. *Journal of Law, Medicine & Ethics, 37*(3), 487–495.

Cling, A. D. (2003). Self-supporting arguments. *Philosophy and Phenomenological Research, 66*(2), 279–303.

Coleman, C. H. (2009). Vulnerability in biomedical research. *Journal of Law, Medicine and Ethics, Spring,* 12–18.

Coleman, C. H., Menikoff, J. A., Goldner, J. A., & Dubler, N. N. (2005). *The ethics and regulation of research with human subjects.* LexisNexis.

Colgrove, N. (2025). Defining 'abortion': A call for clarity. *Theoretical Medicine and Bioethics, 46,* 137–175.

Condic, M. L. (2008). When does human life begin? A scientific perspective. *Westchester Institute White Paper, 1,* 1–31.

Condic, M. L. (2011). A biological definition of the human embryo. In S. Napier (Ed.), *Persons, moral worth, and embryos* (pp. 211–235). Springer Science and Business Media.

Condic, M. L. (2013). When does human life begin: The scientific evidence and terminology revisited. *University of St. Thomas Journal of Law & Public Policy, 8,* 44–81.

Conly, S. (2013). *Against autonomy: Justifying coercive paternalism.* Cambridge University Press.

Conners, S. G. (2012). *Love unleashes life.* Life Cycle Books.

Coss, D. (2018). Interest-relative invariantism and indifference problems. *Acta Analytica, 33*(2), 227–240.

Csikszentmihalyi, M., Abuhamdeh, S., & Nakamura, J. (2014). Flow. In *Flow and the foundations of positive psychology: The collected works of Mihaly Csikszentmihalyi* (pp. 227–238). Springer.

Dancy, R. (1978). On some of Aristotle's second thoughts about substances: Matter. *The Philosophical Review, 87*(3), 372–413.

D'Arcy, E. (1963). *Human acts and their moral evaluation.* Clarendon Press.

Daniels, N. (1979). Wide reflective equilibrium and theory acceptance in ethics. *The Journal of Philosophy, 76*(5), 256–282.

Debes, R. (2009). Dignity's gauntlet. *Philosophical Perspectives, 23*(1), 45–78.

DeGrazia, D. (2005). *Human identity and bioethics.* Cambridge University Press.

DeGrazia, D. (2006). Moral status, human identity, and early embryos: A critique of the president's approach. *The Journal of Law, Medicine & Ethics, 34*(1), 49–57.

Dempsey, M. M. (2012). Victimless conduct and the *volenti* maxim: How consent works. *Criminal Law and Philosophy.* Pub online May 27, 2012. https://doi.org/10.1007/s11572-012-9162-0

DePaul, M. (1993). *Balance and refinement: Beyond coherence methods in ethics.* Routledge.

Derbyshire, S. W. G., & Brockman, J. C. (2020). Reconsidering fetal pain. *Journal of Medical Ethics, 46*(1), 3–6.

DeRose, K. (1992). Contextualism and knowledge attributions. *Philosophy and Phenomenological Research, 52*(4), 913–929.

Diamond, C. (1982). Anything but argument? *Philosophical Investigations, 5*(1), 23–41.

Diamond, C. (1995). *The realistic spirit: Wittgenstein, philosophy, and the mind.* MIT Press.

DiSilvestro, R. (2010). *Human capacities and moral status* (Vol. 108). Springer Science & Business Media.

Doble, J. E., Haig, A. J., Anderson, C., & Katz, R. (2003). Impairment, activity, participation, life satisfaction, and survival in persons with locked-in syndrome for over a decade: Follow-up on a previously reported cohort. *The Journal of Head Trauma Rehabilitation, 18*(5), 435–444.

Dresser, R., & Robertson, J. A. (1989). Quality of life and non-treatment decisions for incompetent patients: A critique of the orthodox approach. *Law, Medicine and Health Care, 17*(3), 234–244.

Dung, L. (2024). Preserving the normative significance of sentience. *Journal of Consciousness Studies, 31*(1-2), 8–30.

Dworkin, R., Nagel, T., Nozick, R., Rawls, J., Scanlon, T. M., & Thomson, J. J. (2012). Assisted suicide: The philosophers' brief. In L. Vaughn (Ed.), *Bioethics: Principles, issues, and cases* (2nd ed.). Oxford University Press.

Eberl, J. T. (2006). *Thomistic principles and bioethics.* Taylor & Francis.

Eberl, J. T. (2010). Fetuses are neither violinists nor violators. *The American Journal of Bioethics, 10*(12), 53–54.

Eberl, J. T. (2020). *The nature of human persons: Metaphysics and bioethics.* University of Notre Dame Pess.

Eberl, J. T., & Brown, B. P. (2011). Brain life and the argument from potential: Affirming the ontological status of human embryos and fetuses. In Persons, moral worth, and embryos. Ed. Stephen Napier (pp. 43–65). Springer Publishers.

Edwards, K., & Smith, E. E. (1996). A disconfirmation bias in the evaluation of arguments. *Journal of Personality and Social Psychology, 71*(1), 5–24.

Elder, C. L. (2005). *Real natures and familiar objects.* The MIT press.

Elga, A. (2007). Reflection and disagreement. *Noûs, 41*(3), 478–502.

Elga, A. (2010). How to disagree about how to disagree. In T. Warfield & R. Feldman (Eds.), *Disagreement* (pp. 175–186). Oxford University Press.

Elliott, C. (1995). Doing harm: Living organ donors, clinical research and the Tenth Man. *Journal of Medical Ethics, 21*, 91–96.

Elliott, C. (2010). *White coat, black hat: Adventures on the dark side of medicine.* Boston, MA.: Beacon Press.

Elliott, C. (2011). *White coat, black hat: Adventures on the dark side of medicine.* Beacon Press.

Eltis, D. (1987). *Economic growth and the ending of the transatlantic slave trade.* Oxford University Press.

Emanuel, E. J., & Miller, F. G. (2007). Money and distorted ethical judgment about research: Ethical assessment of the TeGenero TGN1412 trial. *American Journal of Bioethics, 7*, 76–81.

Emanuel, E. J., Wendler, D., & Grady, C. (2000). What makes clinical research ethical? *JAMA, 283*(20), 2701–2711.

Evans, C. S. (2013). *God and moral obligation.* New York: Oxford University Press.

Fantl, J., & McGrath, M. (2009). *Knowledge in an uncertain world.* Oxford University Press.

Feinberg, J. (1978). Voluntary euthanasia and the inalienable right to life. *Philosophy & Public Affairs, 7*(2), 93–123.

Fergusson, D. M., Horwood, J., & Boden, J. M. (2009). Reactions to abortion and subsequent mental health. *British Journal of Psychiatry, 195*, 420–426.

Finer, L. B., Frohwirth, L. F., Dauphinee, L. A., Singh, S., & Moore, A. M. (2005). Reasons US women have abortions: Quantitative and qualitative perspectives. *Perspectives on Sexual and Reproductive Health, 37*(3), 110–118.

Finn, Peter. (2003). Cannibal case grips Germany; Suspect says internet correspondent volunteered to die. *Washington Post* Dec. 4, at A26.

Finnis, J. (1973). The rights and wrongs of abortion: A reply to Judith Thomson. *Philosophy & Public Affairs, 2*(2), 117–145.

Foot, P. (1967). The problem of abortion and the doctrine of double effect. *Oxford Review, 5*, 5–15.

Forbes, G. (1985). The metaphysics of modality. Clarendon Press.

Forsythe, C. D. (2013). *Abuse of discretion: The inside story of Roe* (Vol. v. Wade). Encounter Books.

Frankl, V. (2010). *The feeling of meaninglessness: a challenge to psychiatry and philosophy.* Marquette University Press.

Freeman, J. B. (2005). *Acceptable premises. An informal approach to an informal logic problem.* Cambridge University Press.

Friberg-Fernros, H. (2017). The moral risk of abortion. In *Making a case for stricter abortion laws* (pp. 21–54). Springer.

Gaita, R. (2004). *Good and evil: An absolute conception*. Routledge.

Gale, R. M. (1972). On what there isn't. *The Review of Metaphysics*, 459–488.

Ganzini, L., & Farrenkopf, T. (1998). Mental health consultation and referral. In K. Haley & M. Lee (Eds.), *The Oregon Death with Dignity Act: A guidebook for health care providers* (pp. 30–32). Oregon Health Sciences University.

Garcia, J. L. A. (1997). Intentions in medical ethics. In J. A. Laing & D. Oderberg (Eds.), *Human lives: Critical essays on consequentialist bioethics* (pp. 161–181). Palgrave Macmillan.

Gawande, A. (2010). Whose body is it anyway? In L. Vaughn (Ed.), *Bioethics: Principles, issues, and cases* (pp. 88–97). Oxford University Press.

George, R. P., & Tollefsen, C. (2011). *Embryo: A defense of human life* (2nd ed.). Witherspoon Institute.

Gerhart, K., Koziol-McLain, J., Lowenstein, S. R., & Whiteneck, G. G. (1994). Quality of life following spinal cord injury: Knowledge and attitudes of emergency care providers. *Annals of Emergency Medicine, 23*(4), 807–812.

Gettier, E. (1963). Is justified true belief knowledge? *Analysis, 23*, 121–123.

Ghorbel, S. (2002). *Statut fonctionnel et qualité de vie chez le locked-in syndrome a domicile*. DEA Motricite Humaine et Handicap, Laboratory of Biostatistics, Epidemiology and Clinical Research. Universite Jean Monnet Saint-Etienne.

Gillham, A. R. (2021). Against the strengthened impairment argument: Never-born fetuses have no FLO to deprive. *Journal of Medical Ethics, 47*(12), e43.

Gill-Thwaites, H. (2006). Lotteries, loopholes and luck: Misdiagnosis in the vegetative state patient. *Brain Injury, 20*(13-14), 1321–1328.

Gilovich, T. (1991). *How we know what isn't so: The fallibility of human reason in everyday life*. Free Press.

Goering, S. (2008). 'You say you're happy, but.': contested quality of life judgments in bioethics and disability studies. *Journal of Bioethical Inquiry, 5*, 125–135.

Goldman, A. (1986). *Epistemology and cognition*. Harvard University Press.

Goldman, A. (2010). The refutation of medical paternalism. In L. Vaughn (Ed.), *Bioethics: Principles, issues, and cases* (pp. 73–78). Oxford University Press.

Goldman, A. I. (1986). *Epistemology and cognition*. Harvard University Press.

Gomez-Lobo, A. (2002). Morality and the human goods: An introduction to natural law ethics. Georgetown University Press.

Gómez-Lobo, A. (2007). Inviolability at any age. *Kennedy Institute of Ethics Journal, 17*(4), 311–320.

Gomez-Lobo, A. and Keown, J. (2015). *Bioethics and the human Goods: an introduction to natural law bioethics*. Washington, D.C.: Georgetown University Press.

Goodin, R. (1985). *Protecting the vulnerable*. University of Chicago Press.

Gorman, M. (2024). A contemporary introduction to Thomistic metaphysics. CUA Press.

Graham, M. (2018). Domains of well-being in minimally conscious patients: Illuminating a persistent problem. *AJOB Neuroscience, 9*(2), 128–130.

Gray, S. (2012). *Love unleashes life: Abortion and the art of communicating truth*. Lifecycle Books.

Greene, G. (1985). *The tenth man*. New York: Simon and Schuster.

Grizes, G. (1970). *Abortion: The myths, the realities, and the arguments*. Corpus books.

Guerrero, A. A. (2007). Don't know, don't kill: Moral ignorance, culpability, and caution. *Philosophical Studies, 136*(1), 59–97.

Habermas, J. (1972). *Knowledge and human interests*. John Wiley & Sons.

Hacker, P. M. S. (2009). *Human nature: The categorial framework*. Wiley-Blackwell.

Hacker, P. M. S. (2013). The intellectual powers: A study of human nature. Hoboken, John Wiley & Sons.

Haidt, J. (2001). The emotional dog and its rational tail: A social intuitionist approach to moral judgment. *Psychological Review, 108*(4), 814–833.

Haidt, J. (2012). *The righteous mind: Why good people are divided by politics and religion*. Pantheon Book.

Haji, I. (2010). Psychopathy, ethical perception, and moral culpability. *Neuroethics, 3*(2), 135–150.

Hamilton, N. G., & Hamilton, C. (2004). Competing paradigms of responding to assisted-suicide requests in Oregon: Case report. http://www.pccef.org/articles/art28.htm. Accessed 18 July 2018.

Hanson, R. (2002). Why health is not special: Errors in evolved bioethics intuitions. *Social Philosophy and Policy, 19*(2), 153–179.

Harman, E. (2015). The irrelevance of moral uncertainty. In R. Shafer-Landau (Ed.), *Oxford studies in metaethics* (Vol. 10, pp. 53–79). Oxford University Press.

Harris, L. H. (2008). Second trimester abortion provision: Breaking the silence and changing the discourse. *Reproductive Health Matters, 16*(31), 74–81.

Hasker, W. (1999). The emergent self. Cornell University Press.

Haslett, D. W. (1987). What is wrong with Reflective Equilibria? *The Philosophical Quarterly, 37*(148), 305–311.

Haybron, D. M. (2008). *The pursuit of unhappiness: The elusive psychology of well-being*. Oxford University Press.

Heaney, S. J. (1992). Aquinas and the presence of the human rational soul in the early embryo. *The Thomist: A Speculative Quarterly Review, 56*(1), 19–48.

Hendricks, P. (2019a). Even if the fetus is not a person, abortion is Immoral: The impairment argument. *Bioethics, 33*(2), 245–253.

Hendricks, P. C. (2019b). (Regrettably) abortion remains immoral: The impairment argument defended. *Bioethics, 33*(8), 968–969.

Henley, K. (1977). The value of individuals. *Philosophy and Phenomenological Research, 37*, 345–352.

Henry, D. (2005). Embryological models in ancient philosophy. *Phronesis, 50*(1), 1–42.

Henry, D. (2008). Organismal natures. *Apeiron, 41*(3), 47–74.

Hershenov, D. (2008). A hylomorphic account of thought experiments concerning personal identity. *American Catholic Philosophical Quarterly, 82*(3), 481–502.

Hershenov, D. (2011). Soulless organisms?: Hylomorphism vs. animalism. *American Catholic Philosophical Quarterly, 85*(3), 465–482.

Hershenov, D. B. (2001). Abortions and distortions: An analysis of morally irrelevant factors in Thomson's violinist thought experiment. *Social Theory and Practice, 27*(1), 129–148.

Hershenov, D. B. (2020). Pathocentric health care and a minimal internal morality of medicine. *Journal of Medicine and Philosophy, 45*(1), 16–27.

Hershenov, D. B. (2023). Two dogmas of pro-life metaphysics. In *Proceedings of the American Catholic Philosophical Association* (Vol. 97, pp. 21–37).

Hershenov, D., & Hershenov, R. (2017). The potential of potentiality arguments. In *Contemporary controversies in Catholic bioethics*. Ed. Jason Eberl. (pp. 35–51). Springer International Publishing.

Hill, B. J. (2010). Abortion as health care. *The American Journal of Bioethics., 10*(12), 48–49.

Holtug, N. (2011). Killing and the time-relative interest account. *The Journal of Ethics, 15*(3), 169–189.

Horton, J., & Ross, J. (2025). Evaluative uncertainty and permissible preference. *Philosophical Review, 134*(1), 35–64.

Howsepian, A. A. (2011). Fetal pains and fetal brains. In S. Napier (Ed.), *Persons, moral worth, and embryos* (Vol. 111, pp. 187–210). Springer Science and Business Media.

Huemer, M. (2011). Epistemological egoism and agent centered norms. In T. Dougherty (Ed.), *Evidentialism and its discontents* (pp. 17–33). Oxford University Press.

Hume, D. (1978). A Treatise of Human Nature (L. A. Selby-Bigge & P. H. Nidditch, Eds.; 2nd ed.). New York: Oxford University Press.

Hurd, H. M. (1996). The moral magic of consent. *Legal Theory, 2*(2), 121–146.

Hursthouse, R. (1991). Virtue theory and abortion. *Philosophy & Public Affairs*, 223–246.

Iyengar, S., & Westwood, S. J. (2015). Fear and loathing across party lines: New evidence on group polarization. *American Journal of Political Science, 59*(3), 690–707.

Jackson, E., & Keown, J. (2012). *Debating euthanasia*. Hart Publishing.

Jansen, L. A. (2010). Disambiguating clinical intentions: The ethics of palliative sedation. *Journal of Medicine and Philosophy, 35*(1), 19–31.

Johnson, A., & Detrow, K. (2016). *The walls are talking: Former abortion clinic workers tell their stories*. Ignatius Press.

Jones, D. A. (2015). Assisted suicide and euthanasia: A guide to the evidence. http://www.bioethics.org.uk/evidenceguide.pdf. Accessed 18 July 2018.

Jonsen, A. R., Siegler, M., & Winslade, W. J. (2002). *Clinical ethics: A practical approach to ethical decisions in clinical medicine* (5th ed.). McGraw-Hill.

Kaake, A. (2013). Pro-choice is pro-violence. ERI Blog. https://blog.equalrightsinstitute.com/pro-choice-is-proviolence/. (Accessed, 1/5/2026).

Kaczor, C. (2011). *The Ethics of abortion: Women's rights, human life and the question of justice*. Routledge.

Kaczor, C. (2013). *A defense of dignity: Creating life, destroying life, and protecting the rights of conscience*. University of Notre Dame Press.

Kaczor, C., & George, R. P. (2017). Death with dignity: A dangerous euphemism. In *Human dignity and assisted death*. Ed. Sebastion Muders. (pp. 68–82). New York: Oxford University Press.

Kaczor, C., & George, R. P. (2018). Death with dignity: A dangerous euphemism. In S. Muders (Ed.), *Human dignity and assisted death* (pp. 68–82). Oxford University Press.

Kagan, S. (2018). The value of life, Part II; Other bad aspects of death, Part I. https://oyc.yale.edu/philosophy/phil-176/lecture-20. Accessed 18 July 2018.

Kaposy, C. (2007). Can infants have interests in continued life? *Theoretical Medicine and Bioethics, 28*(4), 301–330.

Kaposy, C. (2010). Proof and persuasion in the philosophical debate about abortion. *Philosophy & rhetoric, 43*(2), 139–162.

Kappel, K., & Andersen, F. J. (2019). Moral disagreement and higher-order evidence. *Ethical Theory and Moral Practice, 22*(5), 1103–1120.

Kass, L. R. (2008). Defending human dignity. In A. Schulman (Ed.), *Human dignity and bioethics: Essays commissioned by the President's Council on Bioethics* (pp. 297–332). Government printing office.

Kelly, T. (2008). Disagreement, dogmatism, and belief polarization. *Journal of Philosophy, 105*(10), 611–633.

Kelly, T. (2013). Disagreement and the Burdens of Judgment. In D. Christensen & J. Lackey (Eds.), *The epistemology of disagreement: New essays* (pp. 31–53). Oxford University Press.

Keown, J. (2018). *Euthanasia, ethics and public policy: An argument against legalisation*. Cambridge University Press.

King, N. L. (2012). Disagreement: What's the problem? or a good peer is hard to find. *Philosophy and Phenomenological Research., 85*(2), 249–272.

Kohl, M. (1975). Voluntary beneficent euthanasia. In *Beneficent Euthanasia*. Ed. Marvin Kohl (pp. 130–144). Buffalo, NY: Prometheus Books.

Konyndyk, K. J. (1986). Introductory modal logic. University of Notre Dame Press.

Kramer, M. H. (2009). *Moral realism as a moral doctrine* (Vol. 3). Wiley-Blackwell.

Kramer, K., Zaaijer, H. L., & Verweij, M. F. (2017). The precautionary principle and the tolerability of blood transfusion risks. *The American Journal of Bioethics., 17*(3), 32–43.

Kreeft, P. (1990). *Making choices: Practical wisdom for everyday moral decisions*. Servant Books.

Kuhn, T. S. (1996). *The structure of scientific revolutions* (3rd ed.). Chicago University Press.

Kuhse, H., & Singer, P. (2009). In J. P. Lizza (Ed.), *Defining the beginning and end of life: Readings on personal identity and bioethics*. Johns Hopkins University Press.

Kunda, Z. (1999). *Social cognition*. MIT Press.

Lachs, J. (2010). When abstract moralizing runs amok. In L. Vaughn (Ed.), *Bioethics: Principles, issues, and cases* (pp. 561–565). Oxford University Press.

Lackey, J. (2011). A justificationist view of disagreement's epistemic significance. In A. Haddock, A. Millar, & D. Pritchard (Eds.), *Social Epistemology* (pp. 298–325). Oxford University Press.

LaFleur, W. R., Bohme, G., & Shimazono, S. (Eds.). (2008). *Dark medicine: Rationalizing unethical medical research*. Indiana University Press.

LaPointe, T. (2014). Abortion survivor finds doctor who aborted her twin brother – You won't believe what happened. https://www.inquisitr.com/1236243/abortion-survivor-finds-doctor-who-aborted-her-twin-brother-you-wont-believe-what-happened. Accessed 3 July 2024.

Lasonen-Aarnio, M. (2014). Higher-order evidence and the limits of defeat. *Philosophy and Phenomenological Research, 88*(2), 314–345.

Laudan, L. (1990). Demystifying underdetermination. *Minnesota Studies in the Philosophy of Science, 14*(1990), 267–297.

Laureys, S., Pellas, F., van Eeckhout, P., et al. (2005). The locked-in syndrome: What is it like to be conscious but paralyzed and voiceless? In S. Laurey (Ed.), *Progress in brain research* (The boundaries of consciousness: neurobiology and neuropathology) (Vol. 150, pp. 495–511). Elsevier.

Laureys, S., Perrin, F., Schnakers, C., et al. (2005). Residual cognitive function in comatose, vegetative and minimally conscious states. *Current Opinion in Neurology, 18*, 726–733.

Lear, J. (1988). *Aristotle: The desire to understand*. Cambridge University Press.

Lee, P. (2004). A Christian philosopher's view of recent directions in the abortion debate. *Christian Bioethics, 10*, 7–31.

Lee, P. (2010). *Abortion and unborn human life*. CUA Press.

Lee, P., & George, R. P. (2007). *Body-self dualism in contemporary ethics and politics*. New York: Cambridge University Press.

Lee, P., & George, R. P. (2008). The nature and basis of human dignity. *Ratio Juris, 21*(2), 173–193.

Lemos, R. M. (1995). *The nature of value*. University of Florida Press.

Leon-Carrion, J., van Eeckhout, P., & Dominguez-Morales, M. R. (2002). The locked-in syndrome: A syndrome looking for a therapy. *Brain Injury, 16*, 555–569.

Levatino, A. (2016). Abortion procedures: 1st, 2nd, 3rd trimesters. https://www.youtube.com/watch?v=CFZDhM5Gwhk&t=213s

Levy, N. (2024). The value of consciousness. *Journal of Consciousness Studies, 21*(1-2), 127–138.

Lewis, C.I. (1971). *Analysis of knowledge and valuation*. Lasalle, IL.: Open Court Publishing.

Liao, M. (2006). The embryo rescue case. *Theoretical Medicine and Bioethics, 27*(2), 141–147.

Little, M. O. (1995). Seeing and caring: The role of affect in feminist moral epistemology. *Hypatia, 10*(3), 117–137.

Little, M. O. (1998). Cosmetic surgery, suspect norms, and the ethics of complicity. In *Enhancing human traits: Ethical and social implications* (pp. 162–176).

Littlejohn, C. (2013). Disagreement and defeat: Clayton Littlejohn. In D. E. Machuca (Ed.), *Disagreement and skepticism* (pp. 178–201). Routledge.

Liu, Y., Moore, A., Webb, J., & Vallor, S. (2022, July). Artificial moral advisors: A new perspective from moral psychology. In *Proceedings of the 2022 AAAI/ACM conference on AI, ethics, and society* (pp. 436–445).

Lockhart, T. (2000). *Moral uncertainty and its consequences*. Oxford University Press.

Longino, H. E. (1990). *Science as social knowledge: Values and objectivity in scientific inquiry*. Princeton University Press.

Lord, C. G., Ross, L., & Lepper, M. R. (1979). Biased assimilation and attitude polarization: The effects of prior theories on subsequently considered evidence. *Journal of Personality and Social Psychology, 37*(11), 2098–2109.

Loux, M. J. (1998). *Metaphysics: A contemporary introduction*. Routledge.

Lowe, E. J. (1991). Substance and selfhood. *Philosophy, 66*(255), 81–99.

Lu, M. (2011). Abortion and virtue ethics. In S. Napier (Ed.), *Persons, moral worth, and embryos* (pp. 101–123). Springer Nature.

Lu, T. M. (2018). Review of: Kate Greasley and Christopher Kaczor. In *Abortion rights: For and against*. Cambridge University Press. https://ndpr.nd.edu/news/abortion-rights-for-and-against/. Accessed 16 July 2018

Lulé, D., Pauli, S., Altintas, E., et al. (2012). Emotional adjustment in amyotrophic lateral sclerosis (ALS). *Journal of Neurology, 259*(2), 334–341.

Lule, D., Zickler, C., Hacker, S., et al. (2009). Life can be worth living in locked-in syndrome. *Progress in Brain Research, 177,* 339–351.

MacAskill, W., & Ord, T. (2020). Why maximize expected choiceworthiness? *Noûs, 54*(2), 327–353.

Mack, A., & Rock, I. (1998). *Inattentional blindness.* MIT Press.

MacNair, M. R. (2009). *Achieving peace in the abortion war.* iUniverse.

Maitzen, S. (2003). Abortion in the original position. *Personalist Forum, 15*(2), 373–387.

Manfredi, P. L., Morrison, R. S., Morris, J., et al. (2000). Palliative care consultations: How do they impact the care of hospitalized patients? *Journal of Pain and Symptom Management, 20,* 166–173.

Manninen, B. A. (2010). Rethinking Roe v. Wade: Defending the abortion right in the face of contemporary opposition. *The American Journal of Bioethics, 10*(12), 33–46.

Manson, N. C., & O'Neill, O. (2007). *Rethinking informed consent in bioethics.* Cambridge University Press.

Mappes, T. A. (2006). Some reflections on advanced directives. In T. A. Mappes & D. DeGrazia (Eds.), *Biomedical Ethics* (6th ed., pp. 350–357). McGraw-Hill.

Marquis, D. (1989). Why abortion is immoral. *Journal of Philosophy, 86*(4), 183–202.

Marquis, D. (2010). Manninen's defense of abortion rights is unsuccessful. *The American Journal of Bioethics, 10*(12), 56–57.

Martinez-Morales, P. L., Revilla, A., Ocana, I., Gonzalez, C., Sainz, P., McGuire, D., & Liste, I. (2013). Progress in stem cell therapy for major human neurological disorders. *Stem Cell Reviews and Reports., 9*(5), 685–699.

Masek, L. (2009). Intentions, motives and the doctrine of double effect. *The Philosophical Quarterly, 60*(240), 567–585.

May, T. (1998, Fall). Assessing competency without judging merit. *Journal of Clinical Ethics, 9*(3), 247–257.

McDowell, J. (1997). Virtues and reasons. In R. Crisp & M. Slote (Eds.), *Virtue ethics* (Vol. 10, pp. 141–162). Oxford University Press.

McGrath, M. (2021). Undercutting defeat: When it happens and some implications for epistemology. In J. Brown & M. Simion (Eds.), *Reasons, justification, and defeat* (pp. 201–222). Oxford University Press.

McInerney, P. (1990). Does a fetus already have a future like ours? *Journal of Philosophy, 87*(5), 264–268.

McLachlan, H. V. (1977). Must we accept either the conservative or the liberal view on abortion? *Analysis, 37*(4), 197–204.

McMahan, J. (2002). *The ethics of killing: Problems at the margins of life.* Oxford University Press.

McMahan, J. (2006). Paradoxes of abortion and prenatal injury. *Ethics, 116*(4), 625–655.

McMahan, J. (2007). Killing embryos for stem cell research. *Metaphilosophy, 38*(2/3), 170–189.

Meilaender, G. (1984). On removing food and water: Against the stream. *The Hastings Center Report, 14*(6), 11–13.

Meilaender, G. (1998). Terra es animate: On having a life. In A. Verhey & S. Lammers (Eds.), *On Moral Medicine: Theological Perspectives in Medical Ethics* (pp. 390–400). Wm. Eerdman's Publishing.

Melchior, G. (2019). *Knowing and checking: An epistemological investigation.* Routledge.

Miller, F., & Joffe, S. (2009). Limits to research risks. *Journal of Medical Ethics, 35*(7), 445–449.

Miller, G. F., & Wertheimer, A. (2007). Facing up to paternalism in research ethics. *Hastings Center Report, 37*(3), 24–34.

Mills, E. (2008). The egg and I: Conception, identity, and abortion. *Philosophical Review, 117*(3), 323–348.

Moller, D. (2011). Abortion and moral risk1. *Philosophy, 86*(3), 425–443.

Montague, P. (1989). Infant rights and the morality of infanticide. *Nous, 23*(1), 63–81.

Monti, M. M., Vanhaudenhuyse, A., Coleman, M. R., Boly, M., Pickard, J. D., Tshibanda, L., Owen, A. M., & Laureys, S. (2010). Willful modulation of brain activity in disorders of consciousness. *New England Journal of Medicine, 362*(7), 579–589.

Moon, M. (2011). The effects of divorce on children: Married and divorced parents' perspectives. *Journal of Divorce and Remarriage, 52*, 344–349.

Morrison, S. (2010). Palliative care 2020: The future of palliative care. https://www.youtube.com/watch?v=yar-iXPug0A. Accessed 17 May 2018.

Mosteller, T. (2005). Aristotle and headless clones. *Theoretical Medicine and Bioethics, 26*(4), 339–350.

Mullins, A. (2010). The opportunity of adversity. *TED Talks.* http://www.ted.com/talks/aimee_mullins_the_opportunity_of_adversity.html. Accessed 17 January 2014.

Murphy, M. C. (1999). The simple desire-fulfillment theory. *Noûs, 33*(2), 247–272.

Murrell, W., Féron, F., Wetzig, A., Cameron, N., Splatt, K., Bellette, B., et al. (2005). Multipotent stem cells from adult olfactory mucosa. *Developmental Dynamics: An Official Publication of the American Association of Anatomists, 233*(2), 496–515.

Myles-Worsley, M., Johnston, W. A., & Simons, M. A. (1988). The influence of expertise on X-ray image processing. *Journal of Experimental Psychology: Learning, Memory, and Cognition, 14*(3), 553–557.

Nachev, P., & Hacker, P. M. S. (2010). Covert cognition in the persistent vegetative state. *Progress in Neurobiology, 91*(1), 68–76.

Nakase-Richardson, R., Whyte, J., Giacino, J. T., Pavawalla, S., Barnett, S. D., Yablon, S. A., Sherer, M., et al. (2012). Longitudinal outcome of patients with disordered consciousness in the NIDRR TBI Model Systems Programs. *Journal of Neurotrauma, 29*(1), 59–65.

Napier, S. (2008). *Virtue epistemology: Motivation and knowledge.* Bloomsbury.

Napier, S. (2010). Vulnerable embryos: A critical analysis of twinning, rescue, and natural-loss arguments. *American Catholic Philosophical Quarterly, 84*(4), 781–810.

Napier, S. (2013). Challenging research on human subjects: Justice and uncompensated harms. *Theoretical Medicine and Bioethics, 34*, 29.

Napier, S. (2015). The justification of killing and psychological accounts of the person. *American Catholic Philosophical Quarterly, 89*(4), 651–680.

Napier, S. (2016). Thought experiments, the reliability of intuitions, and human embryonic stem cell research. *International Philosophical Quarterly, 56*(1), 77–99.

Napier, S. (2020). *Uncertain bioethics: Moral risk and human dignity.* Routledge.

Napier, S. (2024). Impairment Arguments, Interests, and Circularity. *Journal of Medicine and Philosophy, 49*(5), 470–480.

Napier, S. (n.d.). Conscience rights, naturalism, and moral knowledge: In defense of conscience absolutism. *Journal of Medicine and Philosophy.* (in press).

Nettleman, M., Ingersoll, K. S., & Ceperich, S. D. (2006). Characteristics of adult women who abstain from sexual intercourse. *BMJ Sexual & Reproductive Health, 32*(1), 23–24.

Nichols, P. (2012). Abortion, time-relative interests, and futures like ours. *Ethical Theory and Moral Practice, 15*(4), 493–506.

Noonan, H. W. (1985). The closest continuer theory of identity. *Inquiry, 28*(1-4), 195–229.

Norton, S. A., Hogan, L. A., Holloway, R. G., Temkin-Greener, H., Buckley, M. J., & Quill, T. E. (2007). Proactive palliative care in the medical intensive care unit: Effects on length of stay for selected high-risk patients. *Critical Care Medicine, 35*(6), 1530–1535.

Not Dead Yet. (2018). Not dead yet disability activists oppose assisted suicide as a deadly form of discrimination. http://notdeadyet.org/assisted-suicide-talking-points. Accessed 18 July 2018.

Oderberg, D. (2008). *Real essentialism.* Routledge.

Oderberg, D. (n.d.). The origination of a human being: Rejoinder to person (unpublished). https://drive.google.com/file/d/0B7SKlRTfkUiecmdYN2lRQm8zdFE/view. Accessed 18 July 2018.

Oderberg, D. S. (1997). Voluntary euthanasia and justice. In J. Laing & D. Oderberg (Eds.), *Human lives: Critical essays in consequentialist bioethics* (pp. 225–240). Palgrave Macmillan.

Olson, E. T. (1997). *The human animal: Personal identity without psychology.* Oxford University Press.

Olson, E. T. (2002). Thinking animals and the reference of 'I'. *Philosophical Topics, 30*(1), 189–207.

Olson, E. T. (2003). An argument for animalism. In R. Martin & J. Barresi (Eds.), *Personal identity* (pp. 318–334). Blackwell Publishing.

Olson, E. T. (2007). *What are we?: A study in personal ontology.* Oxford University Press.

Olson, E. T. (2017). Personal identity. In E. N. Zalta (Ed.), *The Stanford encyclopedia of philosophy* (Summer 2017 edition) https://plato.stanford.edu/archives/sum2017/entries/identity-personal/

Oregon Department of Health. (2014). Oregon death with Dignity Act – 2014. DWD2014ARnarrative20150202x.

Oregon Health Division. (2004). *Sixth annual report on Oregon's Death with Dignity Act.* http://www.ohd.hr.state.or.us/chs/pas/ar-index.cfm

Otte, R. (n.d.). Science, naturalism, and self defeat (manuscript). https://people.ucsc.edu/~otte/articles/nat.china.pdf. Accessed 30 July 2024.

Parfit, D. (1971). Personal identity. *The Philosophical Review, 80*(1), 3–27.

Pasnau, R. (2014). Veiled disagreement. *The Journal of Philosophy, 111*(11), 608–630.

Patel, C., Kleinig, P., Bakker, M., & Tait, P. (2019). Palliative sedation: 'A safety net for the relief of refractory and intolerable symptoms at the end of life'. *Australian Journal of General Practice, 48*(12), 838–845.

Pellegrino, E. (1996). The place of intention in the moral assessment of assisted suicide and active euthanasia. In T. Beauchamp (Ed.), *Intending death: The ethics of assisted suicide and euthanasia* (pp. 163–183). Prentice Hall.

Pellegrino, E., & Thomasma, D. (1988). *For the patient's good: Restoration of beneficence in health care.* Oxford University Press.

Persson, I. (1999). Harming the non-conscious. *Bioethics, 13*(3-4), 294–305.

Pierson, L., Gibert, S., Orszag, L., Sullivan, H. K., Fei, R. Y., Persad, G., & Largent, E. A. (2024). Bioethicists today: Results of the views in bioethics survey. *The American Journal of Bioethics, 24*(9), 9–24.

Pilsner, J. (2006). *The specification of human actions in St Thomas Aquinas.* Oxford University Press.

Pizarro, D. (2000). Nothing more than feelings? The role of emotions in moral judgment. *Journal for the theory of social behaviour, 30*(3), 355–375.

Plantinga, A. (1974). *The nature of necessity.* Oxford University Press.

Plantinga, A. (1993). *Warrant: The current debate.* Oxford University Press.

Plantinga, A. (2000). *Warranted christian belief.* Oxford University Press.

Pollock, J., & Cruz, J. (1999). *Contemporary theories of knowledge* (2nd ed.). Rowman and Littlefield.

Power, M. (2013). Confessions of a drone warrior. GQ [On-line]. https://www.gq.com/story/drone-uav-pilot-assassination. Accessed 11 August 2025.

Prusak, B. G. (2013). *Parental obligations and bioethics: The duties of a creator.* Routledge.

Pruss, A. (2011). I was once a fetus: That is why abortion is wrong. In Persons, moral worth, and embryos: A critical analysis of pro-choice arguments ed. Stephen Napier (pp. 19–42). Springer Netherlands.

Pruss, A. (2013). Aristotelian forms and laws of nature. *Analysis and Existence, 24*, 115–132.

Pruss, A. (2017). Being a bad person and doing wrong. http://alexanderpruss.blogspot.com/2017/12/being-bad-person-and-doing-wrong.html. Accessed 15 March 2018.

Puccetti, R. (1983). The life of a person. In W. B. Bondesson, H. T. Englehardt, S. Spicker, & D. H. Winship (Eds.), *Abortion and the status of the fetus.* D. Reidel Publishing Company.

Rachels, J. (1986). *The end of life: Euthanasia and morality.* Oxford University Press.

Reed, B. (2012). Resisting encroachment. *Philosophy and Phenomenological Research, 85*(2), 465–472.

Rescher, N. (2006). *Presumption and the practices of tentative cognition.* Cambridge University Press.

Richardson, R. C. (2000). The organism in development. *Philosophy of Science, 67*, S312–S321.

Roberts, R. C., & Wood, W. J. (2008). *Intellectual virtues: An essay in regulative epistemology.* Oxford University Press.

Roe, K. M. (1989). Private troubles and public issues: Providing abortion amid competing definitions. *Social Science & Medicine, 29*(10), 1191–1198.

Rosen, M. (2018). *Dignity: Its history and meaning.* Harvard University Press.

Royal College of Psychiatrists. (2006). *Statement from the Royal College of Psychiatrists on Physician-Assisted Suicide.* https://www.rcpsych.ac.uk/pdf/CPC%2004.05.11%20Enc%2010. pdf. Accessed 11 May 2018.

Rudner, R. (1953). Scientist qua Scientist makes value judgments. *Philosophy of Science, 20*(1), 1–6.

Ruminjo, A., & Mekinulov, B. (2008 Jun). A case report of Cotard's Syndrome. *Psychiatry (Edgmont), 5*(6), 28–29.

Rutiku R, Aru J and Bachmann T. (2016). General Markers of Conscious Visual Perception and Their Timing. *Frontiers in Human Neuroscience, 10*, 23. https://doi.org/10.3389/fnhum.2016.00023.

Salamanca-Balen, N., Merluzzi, T. V., & Chen, M. (2021). The effectiveness of hope-fostering interventions in palliative care: A systematic review and meta-analysis. *Palliative Medicine, 35*(4), 710–728.

Sandel, M. J. (2005). The ethical implications of human cloning. *Perspectives in Biology and Medicine, 48*(2005), 241–246.

Savulescu, J., & Schuklenk, U. (2017). Doctors have no right to refuse medical assistance in dying, abortion or contraception. *Bioethics, 31*(3), 162–170.

Scaltsas, T. (1994). Substantial holism. In T. Scaltsas, D. Charles, & M. L. Gill (Eds.), *Unity, identity, and explanation in Aristotle's metaphysics* (pp. 107–128). Oxford University Press.

Schwarz, S. D. (1990). *The moral question of abortion.* Loyola University Press.

Schwarz, S. D., & Tacelli, R. K. (1989). Abortion and some philosophers: A critical examination. *Public Affairs Quarterly, 3*(2), 81–98.

Seidel, M. (2014). *Epistemic relativism: A constructive critique.* Springer.

Sepielli, A. (2019). Decision making under moral uncertainty. In A. Zimmerman, K. Jones, & M. Timmons (Eds.), *The Routledge handbook of moral epistemology* (pp. 508–521). Routledge.

Sgaravatti, D. (2013). Petitio principii: A bad form of reasoning. *Mind, 122*(487), 749–779.

Shaw, J. (2006). Intention in ethics. *Canadian Journal of Philosophy, 36*(2), 187–223.

Shaw, J. (2015). Death and other harms: Intention and the problem of closeness. *American Catholic Philosophical Quarterly, 89*(3), 421–439.

Sherman, N. (1991). *The fabric of character: Aristotle's theory of virtue.* Oxford University Press.

Shoemaker, S. (1963). *Self-knowledge and self-identity.* Cornell University Press.

Shoemaker, S. (1984). Personal identity. A materialist's account. In S. Shoemaker & R. Swinburne (Eds.), *Personal identity* (pp. 67–132). Oxford University Press.

Shoemaker, S. (1999). Self, body, and coincidence. *Aristotelian Society Supplementary, 73*, 287–306. https://doi.org/10.1111/1467-8349.00059

Shoemaker, S. (2016). Thinking animals without animalism. In S. Blatti & P. Snowdon (Eds.), *Animalism: New essays on persons, animals, and identity* (pp. 128–142). Oxford University Press.

Singer, P. (1979). *Practical ethics.* Cambridge University Press.

Sloan, D., & Hartz, P. (1992). *Abortion: Doctor's perspective, a woman's dilemma.* Donald I Fine.

Smith, S. (2012). Sarah Smith. https://clinicquotes.com/category/articles/abortion-survivors/page/2/. Accessed 27 May 2025.

Snowdon, P. F. (2014). *Persons, animals, ourselves.* Oxford University Press.

Soulier, M., Maislen, J. D., & Beck, J. (2010). Status of the psychiatric duty to protect, circa 2006. *Journal of the American Acadamy of Psychiatry and Law, 38*, 457–473.

Sprague, R. K. (1990). Aristotle on mutilation: Metaphysics 5.27. *Syllecta Classica, 2*(1), 17–22.

Stanford, K. (2017). Underdetermination of scientific theory. In E. N. Zalta (Ed.), *The Stanford encyclopedia of philosophy* (Winter 2017 edition) https://plato.stanford.edu/archives/win2017/entries/scientific-underdetermination/

Stanley, J. (2005). *Knowledge and practical interests.* Oxford University Press.

Statement from the Royal College of Psychiatrists on Physician-Assisted Suicide. (2006). http://www.rcpsych.ac.uk/pressparliament/collegeresponses/physicianassistedsuicide.aspx. para 2.4. (Original emphasis. Last accessed 10 September 2007).

Stickgold, R. (2005). Sleep-dependent memory consolidation. *Nature, 437*(7063), 1272–1278.

Stith, R. (2014). Construction vs. development: Polarizing models of human gestation. *Kennedy Institute of Ethics Journal, 24*(4), 345–384.

Strawson, G. (2003). The self. In R. Martin & J. Barresi (Eds.), *Personal identity* (pp. 335–377). Blackwell Publishing.

Stretton, D. (2004). Essential properties and the right to life: A response to Lee. *Bioethics, 18*(3), 264–282.

Stretton, D. (2008). Critical notice—Defending life: A moral and legal case against abortion choice by Francis J Beckwith. *Journal of Medical Ethics, 34*(11), 793–797.

Stuchlik, J. (2021). *Intention and wrongdoing: in defense of double effect.* New York: Cambridge University Press.

Stump, E. (2010). *Wandering in darkness: Narrative and the problem of suffering.* Oxford University Press.

Sturgeon, S. (2014). Pollock on defeasible reasons. *Philosophical Studies, 169*, 105–118.

Sulmasy, D. P. (2008). Dignity and bioethics: History, theory, and selected applications. In A. Schulman (Ed.), *Human dignity and bioethics: Essays commissioned by the President's Council on Bioethics* (pp. 469–504). Government printing office.

Sulmasy, D. P. (2018). Sedation and care at the end of life. *Theoretical Medicine and Bioethics, 39*(3), 171–180.

Sumner, L. W. (2011). *Assisted death: a study in ethics and law.* New York: Oxford University Press.

Swain, M. (1981). *Reasons and knowledge.* Cornell University Press.

Taboada, P. (Ed.). (2015). *Sedation at the End-of-life: An Interdisciplinary Approach.* Springer Publishers.

Tadros, V. (2011). Consent to harm. *Current Legal Problems, 64*, 23–49.

Taylor, J. S. (2016). Autonomy, competence, and end of life. In J. K. Davis (Ed.), *Ethics at the end of life: New issues and arguments* (pp. 101–115). Routledge.

Thomas, S. (2021). *The rights and wrongs of abortion: A philosophy and public affairs reader.* Princeton University Press.

Thomas, C., Alici, Y., Breitbart, W., Bruera, E., Blackler, L., & Sulmasy, D. P. (2024). Drugs, delirium, and ethics at the end of life. *Journal of the American Geriatrics Society, 72*(7), 1964–1972. https://doi.org/10.1111/jgs.18766

Thomson, J. J. (1971). A defense of abortion. *Philosophy & Public Affairs, 1*(1), 47–66.

Thomson, J. J. (1995). Abortion. *Boston Review* (online), pp. 11–15 (print). http://bostonreview.net/archives/BR20.3/thomson.html. Accessed 18 July 2018.

Thune, M. (2010). 'Partial defeaters' and the epistemology of disagreement. *The Philosophical Quarterly, 60*(239), 355–372.

Tiozzo, M. (2024). Dualism about undercutting defeat. *Ratio, 37*(2–3), 145–155.

Tollefsen, C. (2006). Is a purely first-person account of human action defensible? *Ethical Theory and Moral Practice, 9*(4), 441–460.

Tollefsen, C. (2010). Fetal interests, fetal persons, and human goods. In S. Napier (Ed.), *Persons, moral worth, and embryos: A critical analysis of pro-choice arguments* (pp. 163–184). Springer Science and Business Media.

Tollefsen, C. (2011). Fetal interests, fetal persons, and human goods. In *Persons, moral worth, and embryos.* Ed. Stephen Napier (pp. 163–184). Springer Publishers.

Tooley, M. (1972). Abortion and infanticide. *Philosophy & Public Affairs., 2*(1), 37–65.

Tooley, M. (1983). Abortion and infanticide. Oxford University Press.

Trotter, G. (2010). Abortion, secular dogma, and the sacrament of sex: Another failed attempt to impose moral idiosyncrasies through the ruse of argument. *The American Journal of Bioethics, 10*(12), 51–52.

Vuksanovic, D., Green, H. J., Dyck, M., & Morrissey, S. A. (2017). Dignity therapy and life review for palliative care patients: A randomized controlled trial. *Journal of Pain and Symptom Management, 53*(2), 162–170.

Walton, D. N. (1985). Are circular arguments necessarily vicious? *American Philosophical Quarterly, 22*(4), 263–274.

Walton, D. N. (1991). *Begging the question: Circular reasoning as a tactic of argumentation.* Greenwood Press.

Warren, M. A. (1973). On the moral and legal status of abortion. *The Monist, 57*(1), 43–61.

Warren, M. A. (1992). *Moral status: Obligations to persons and other living things.* Clarendon Press.

Watt, H. (2000). *Life and death in healthcare ethics: A short introduction.* Routledge.

Watt, H. (2022). Do fetuses have the same interests as their mothers? In N. Colgrove, B. P. Blackshaw, & D. Rodger (Eds.), *Agency, pregnancy and persons* (pp. 105–123). Routledge.

Weatherson, B. (2014). Running risks morally. *Philosophical Studies, 167*(1), 141–163.

Weil, S. (1968). The love of god and affliction. In S. Weil (Ed.), *Science, necessity and the love of God.* Oxford University Press.

Weijer, C. (2000). The ethical analysis of risk. *Journal of Law, Medicine & Ethics, 28*(4), 344–361.

Wellman, C. (1971). *Challenge and response: Justification in ethics.* Southern Illinois University Press.

Wertheimer, R. (1971). Understanding the abortion argument. *Philosophy & Public Affairs, 1*(1), 67–95.

Whitby, B. (2008). Computing machinery and morality. *AI & Society, 22*(4), 551–563.

Wiggins, D. (1974). Essentialism, continuity, and identity. *Synthese, 28*(3-4), 321–359.

Wiggins, D. (2001). *Sameness and substance renewed.* Cambridge University Press.

Williams, B. (1985). *Ethics and the Limits of Philosophy.* Harvard University Press.

Willke, J. C. (1984). *Abortion and slavery: History repeats.* Hayes.

Wilson, R. F. (2009). Estate of Gelsinger v. Trustees of University of Pennsylvania: Money, prestige, and conflicts of interest in human subjects research. *Health Law and Bioethics: Cases in Context,* 229.

Wipond, R. (2024). Dramatic rise in police interventions on 988 callers. https://www.madinamerica.com/2024/06/dramatic-rise-in-police-interventions-on-988-callers/. Accessed 31 July 2024.

Wolterstorff, N. (2008). *Justice: Rgiths and wrongs.* Princeton NJ.: Princeton University Press.

Wreen, M. (1988). The definition of euthanasia. *Philosophy and Phenomenological Research, 48*(4), 637–653.

Wreen, M. J. (2004). Hypothetical autonomy and actual autonomy: Some problem cases involving advance directives. *The Journal of Clinical Ethics, 15*(4), 319–333.

Young, R. (2018). Voluntary euthanasia. In E. N. Zalta (Ed.), *The Stanford encyclopedia of philosophy* (Summer 2018 Edition) https://plato.stanford.edu/archives/sum2018/entries/euthanasia-voluntary/

Young, R. (2022). Voluntary Euthanasia. *The Stanford Encyclopedia of Philosophy.* Edward N. Zalta (ed.), URL = https://plato.stanford.edu/archives/sum2022/entries/euthanasia-voluntary/.

Zagzebski, L. (2001). The uniqueness of persons. *Journal of Religious Ethics, 29*(3), 401–423.

Zagzebski, L. (2017). *Exemplarist moral theory.* New York: Oxford University Press.

Zagzebski, L. T. (1996). *Virtues of the mind: An inquiry into the nature of virtue and the ethical foundations of knowledge.* Cambridge University Press.

Zagzebski, L. T. (2004). *Divine motivation theory.* New York: Cambridge University Press.

Zagzebski, T. L. (2012). *Epistemic authority: A theory of trust, authority and autonomy in belief.* Oxford University Press.

Zuniga, G. (2004). An ontology of dignity. *Metaphysica, 5*(2), 115–131.

Index

© The Author(s), under exclusive license to Springer Nature Switzerland AG 2026
S. Napier, *Justified Killing*, https://doi.org/10.1007/978-3-032-14946-6

GPSR Compliance
The European Union's (EU) General Product Safety Regulation (GPSR) is a set
of rules that requires consumer products to be safe and our obligations to
ensure this.

If you have any concerns about our products, you can contact us on

ProductSafety@springernature.com

In case Publisher is established outside the EU, the EU authorized
representative is:

Springer Nature Customer Service Center GmbH
Europaplatz 3
69115 Heidelberg, Germany

www.ingramcontent.com/pod-product-compliance
Ingram Content Group UK Ltd.
Pitfield, Milton Keynes, MK11 3LW, UK
UKHW021016080726
473054UK00003B/155